Science and Fiction

For further volumes:
http://www.springer.com/series/11657

Science and Fiction – A Springer Series

This collection of entertaining and thought-provoking books will appeal equally to science buffs, scientists and science-fiction fans. It was born out of the recognition that scientific discovery and the creation of plausible fictional scenarios are often two sides of the same coin. Each relies on an understanding of the way the world works, coupled with the imaginative ability to invent new or alternative explanations - and even other worlds. Authored by practicing scientists as well as writers of hard science fiction, these books explore and exploit the borderlands between accepted science and its fictional counterpart. Uncovering mutual influences, promoting fruitful interaction, narrating and analyzing fictional scenarios, together they serve as a reaction vessel for inspired new ideas in science, technology, and beyond.

Whether fiction, fact, or forever undecidable: the Springer Series "Science and Fiction" intends to go where no one has gone before!

Its largely non-technical books take several different approaches. Journey with their authors as they

* Indulge in science speculation – describing intriguing, plausible yet unproven ideas;
* Exploit science fiction for educational purposes and as a means of promoting critical thinking;
* Explore the interplay of science and science fiction – throughout the history of the genre and looking ahead;
* Delve into related topics including, but not limited to: science as a creative process, the limits of science, interplay of literature and knowledge;
* Tell fictional short stories built around well-defined scientific ideas, with a supplement summarizing the science underlying the plot.

Readers can look forward to a broad range of topics, as intriguing as they are important. Here just a few by way of illustration:

* Time travel, superluminal travel, wormholes, teleportation
* Extraterrestrial intelligence and alien civilizations
* Artificial intelligence, planetary brains, the universe as a computer, simulated worlds
* Non-anthropocentric viewpoints
* Synthetic biology, genetic engineering, developing nanotechnologies
* Eco/infrastructure/meteorite-impact disaster scenarios
* Future scenarios, transhumanism, posthumanism, intelligence explosion
* Virtual worlds, cyberspace dramas
* Consciousness and mind manipulation

Nick Kanas

The Caloris Network

A Scientific Novel

 Springer

Nick Kanas
University of California, San Francisco (UCSF)
San Francisco
California
USA

Additional material to this book can be downloaded from https://nickkanas.com/

The persons, characters, events and firms depicted in the fictional part of this work are fictitious. No similarity to actual persons, living or dead, or to actual events or firms is intended or should be inferred. While the advice and information in the science part of this work are believed to be true and accurate at the date of publication, neither the authors nor the editors nor the publisher can accept any legal responsibility for any errors or omissions that may be made. The publisher makes no warranty, express or implied, and accepts no liability with respect to the material contained in either science or fiction parts of the work.

ISSN 2197-1188   ISSN 2197-1196   (electronic)
Science and Fiction
ISBN 978-3-319-30577-6   ISBN 978-3-319-30579-0   (eBook)
DOI 10.1007/978-3-319-30579-0

Library of Congress Control Number: 2016939485

Cover illustration: Man walks through the fantasy crystal corridor with rugged walls and bright glowing end. By Eugene Sergeev/Shutterstock.com.

Printed on acid-free paper

This Springer imprint is published by Springer Nature
The registered company is Springer International Publishing AG Switzerland

Preface

Mercury is the innermost planet of our Solar System. It is around 58 million kilometers (36 million miles) from the Sun, and its daytime temperatures can go up to 427 °C (801 °F). At the other extreme, its nighttime temperatures can drop to −173 °C (−279 °F). Although there is evidence that ice exists in craters located near the more temperate north and south poles, the rest of the planet is waterless and desolate.

I woke up one morning thinking about what kind of life might develop in such an environment! This challenge formed the germ of this novel, and I hope that you will enjoy reading it.

In writing *The Caloris Network*, I want to thank a number of individuals whose help and influence contributed to its final publication. First and foremost is Dr. Christian Caron, the Editor of Springer's "Science and Fiction" series. Chris has selected two of my other novels for publication in this excellent series: *The New Martians* and *The Protos Mandate*. I am grateful to him for his helpful comments on an earlier draft of this novel, along with the comments made by an anonymous member of the series' editorial board. I am also grateful for the useful comments made to an earlier draft by a number of friends and colleagues: Drs. John Holzrichter, Shirley Huang, Lyn Motai, and Richard Ray. Last but not least, I am grateful to my wife Carolynn, who has read and commented on many of my science fiction writings and who has continued to support me in this and many other activities over the years. Of course, I am solely responsible for the ideas and concepts that appear in this book.

This book is dedicated to the memory of my parents, Andrew and Angeline Kanas, whose spirits inhabit the storyline in many ways.

San Francisco, California, USA
January 2016

Nick Kanas
Website: https://nickkanas.com/

Contents

Part I

The Novel

The area around Caloris Basin on Mercury, as photographed by the Mariner 10 spacecraft on March 29, 1974. Note the Basin in the left half of the image and the mountains circling concentrically around it. These features were caused by the collision of a large body with Mercury eons ago. Courtesy of NASA (NASA/NSSDC digital image); and *Solar System Maps: From Antiquity to the Space Age*, Nick Kanas, Springer/Praxis, 2014

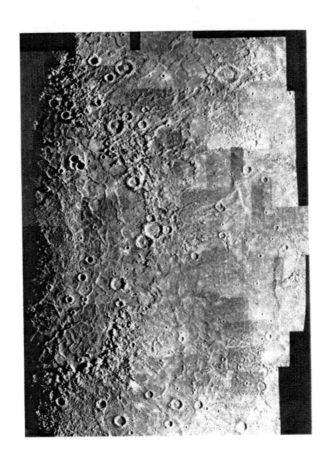

The Caloris Network

Prologue: Mercury, Day 43, 2130

The Sun had stopped in the sky. Soon it would move east for a couple of days, pause again, then continue its trek toward the western horizon, where it would set.

Only on Mercury, Samantha Evans thought, trying to get her mind off of the events that took place earlier in the day.

The actions of the Sun were all a matter of perspective. Mercury's rotational period was tidally coupled to its orbital period in a 3:2 resonance, so it rotated one and a half times during each passage around its bright star. This meant that at local noon on the surface of the planet, the Sun would appear to move back in its path across the sky before continuing its westerly motion.

Samantha put on her filtered glasses and looked directly at the Sun. Appearing more than three times larger than seen from Earth, the chromosphere glowed softly in crimson. She saw a band of sunspots that were moving across the boiling surface. Prominences shot out into space. The Sun looked alive.

Glancing at the outside temperature gauge, she noted that the surface temperature had risen to 310 °C. The heat scorched the cracked surface of the planet with ferocious intensity. Crevices and craterlets dotted the ground in all directions. Four weeks from now, as the Sun continued to heat the surface, the temperature would rise to over 400°, well past the melting point of lead. Mercury indeed was an alien place.

And then there was the matter of the two bodies they were bringing back to the base and her conclusions about the network. . .

How do I tell them? she thought. *Will they believe me? Will they think I'm crazy?*

She wearily rubbed her eyes and glanced over at Chuck Morgan as he steered *Rover 1* across the bumpy surface. His gray eyes darted left and right to avoid the small ridges and potholes in their path. Hair stubble growing on his

© Springer International Publishing Switzerland 2016
N. Kanas, *The Caloris Network*, Science and Fiction, DOI 10.1007/978-3-319-30579-0_1

previously shaved head and face gave him an unkempt appearance. Ahead, the *Hermes* was silhouetted like a sleek silver bullet against the mountains toward the east.

Chuck reached over to activate the comm.

"Hello, Tilda. We're four clicks away and should be home in just a few minutes."

"Roger that," came the static-modulated response from the communications officer. "I have you on the screen. Welcome back. You've had a tough trip."

The 46-year-old muscular engineer frowned.

"That's for damn sure. I imagine the Captain will be holding an emergency briefing."

"That's already planned. Drive safely. Chang out."

Chuck turned the comm off and glanced over at Samantha.

"What do you think will happen?" he asked.

"I don't know. Captain Kilborn will be by-the-book, as usual. I'm sure that after our briefing, he'll contact Luna City. Then, who knows what we'll do."

They drove on in silence. Samantha thought about the network again. This led her to reminisce about her father and what had happened to him. Were there some clues there that related to the network and to herself? Her thoughts became lost in reverie. . .

Station Delta, 28 Years Earlier

"Samantha, hurry up and put your toys away. Your father will be home soon."

"Yes, mother," the 6-year-old said. She glanced at the digital clock on the wall. It read: "18:34/Nov12/2102".

In their small apartment, she didn't have far to go to reach the toys scattered here and there. Besides a small bedroom for her parents and a tiny bathroom/shower, there was only a combined kitchen-living area where the family spent most of their time. Her sleeping space was recessed in one of the walls, which was now closed off by the pulled-up bed. The furniture was minimalist, and two walls were covered by pictures and an entertainment section with a large 3-D holographic projector and multiphonic sound system. On the only outside wall there was a single large view port that opened up to the vastness of space. The kitchen area was dominated by the foodbot, which prepared and presented their meals through its delivery chute.

Samantha reluctantly left her spot by the window to gather her toys. She had been mesmerized by the stunning view of the Earth below, still glorious despite being blanketed in a brownish haze.

Sixty years earlier, overpopulation and climate change had led to a crash program to venture into space in order to help deal with the effects of global warming. Too many people were using too many resources and producing too much waste. Although conservation programs had helped to stabilize things on Earth, it would be decades before the effects of climate change could be reversed. Consequently, plans were made to construct a series of orbiting space stations and bases on the Moon, Mars, and elsewhere in the Solar System to offload as many people as possible.

Station Delta was the fourth such facility orbiting the Earth that was built for this purpose. The giant four-spoked rotating wheel had become a city of over 100,000 people poised to celebrate their 40th Anniversary in the next year. As a senior electrician, her father was working overtime in wiring the float that Sector P would be submitting for the parade through the Main Corridor.

As she picked up her dolls and computer pad, Samantha glanced over at her mother. The slim, tall black woman was at the foodbot, punching the buttons that selected their dinner. She had picked up her daughter from school after her day's work as a nutritionist in the algae farms, and the two of them had only just returned home. As usual, their day had been busy and full of multitasking.

There was a rustling outside. The door dilated open on overlapping metal plates like the iris of an eye, revealing her father. Stocky, with fair skin and blondish hair, he initially looked tired, but upon seeing his family, his face broke out into its characteristic wide, jovial smile.

"Oh, you beat me home," he said. He went over and gave his wife a kiss, then turned toward his daughter with open arms.

"And how is Daddy's little girl?"

Samantha giggled and leaped over toward him, giving him a hug.

"Fine, Daddy," she said.

"Here's a present for you."

He reached into the bag slung over his left shoulder and produced a tiny brown-colored rose.

"I passed the hydroponic gardens on my way home and saw this. It reminded me of you."

He gave it to his olive-complected daughter, whose brown eyes shown with delight.

"Thank you, Daddy," she squealed and excitedly hugged him.

"That's lovely, Dylan," her mother said.

"And I didn't forget you, Kalisha" he responded.

He again reached into his bag and took out a stunning black rose, almost glowing in its iridescent shimmer.

"Thank you, sweetheart," she responded, taking the flower and then pushing another button on the foodbot. "I'm going to order you a second portion of blackberry pie for dessert. And then maybe we'll have our own special dessert after dinner."

They both giggled.

Observing them, and based on past experience, Samantha knew that she would be sent to her bed early tonight.

"How was your day," he asked his wife.

"Pretty good," she responded. "The algae are healthy, and the polls show that our new taste supplements have been well-received by the study group."

"I certainly agree with that!" he responded. "Who would have thought that algae could be made into a patty that tastes just like rib-eye steak? You and your teammates are really clever."

"Necessity is the mother of invention. How is the float coming along?"

"It should be spectacular. The front carries a model of a space ship bound for Saturn. Behind this is a carriage depicting the surface of Titan that shows an ethane pool containing life forms and a mine being worked on by two space-suited miners. The space ship has lights that flash on and off, and the miners move. It's really cool. Our sector should win one of the awards at the celebration next year."

"Just make sure that it doesn't break down and block the other floats behind it, like what happened with the Sector K float 4 years ago."

"Right. There's barely enough room in the Main Corridor for the floats to parade as it is. But I think that our float will function normally and get through just fine."

"Good," Kalisha said, as the foodbot gave a ring. "Whoa, your algae steaks are ready!"

Both Samantha and her father rushed to the small table in anticipation of their dinner.

Mercury, Day 1

"Prepare to land!" Bill Kilborn exclaimed.

Below them the surface was dark, although the Sun could be seen above the horizon at their height. As they descended, the fiery solar ball slipped below the eastern rim of the basin. Caloris Basin was huge—some 1550 km across, formed by a gigantic impact event three and a half billion years ago. It was also the point where the Sun was directly overhead at Mercury's closest approach, making the basin very hot; hence, its name.

"Mr. Brown, fire the retrorockets and ease this baby down," Kilborn commanded.

The pilot undertook a series of deft maneuvers. The giant ship slowly descended toward the ground, the landing lights illuminating the spot that would be their home during their stay on the planet. As the ship approached the surface, support struts extended down and out from the lower hull to stabilize the ship in its vertical orientation. Strapped in their padded acceleration couches, the crewmembers anxiously awaited the landing. It was indeed a momentous event.

Mercury was the last of the Solar System planets to be explored by a landing party. When the first probes reached the scorched planet in the 1970s, it looked desolate. Although subsequent flybys found water ice in craters near the cooler poles, there was no liquid water anywhere, and consequently no sign of carbon-based life. Attention turned to the other planets. Mars was explored in the late 2030s, and methane-producing microbes were found in hot spring caverns in the volcanic Tharsis region. Resource and climate change concerns on Earth pushed governments and entrepreneurs to fund ventures to the asteroids and the moons of Jupiter and Saturn in the latter part of the twenty-first Century in order to look for raw materials that could be mined. These activities yielded a bonus when primitive carbon-based lifeforms were discovered in the subsurface waters of Jupiter's moon Europa and the methane and ethane pools of Titan. The moons of Uranus and Neptune subsequently were explored by probes as possible sites for mines and colonies as the Earth smothered and roasted during the Great Pollution.

Then, in 2127, a probe bound for the Kuiper Belt using the Sun as a gravity assist picked up an electromagnetic energy source from Mercury that was centered in Caloris Basin. Other probes flying by sometimes received a similar EM pulse and sometimes didn't, especially when the Basin was in nighttime darkness. The media had a field day with the phenomenon, some reports stating that this was a signal that suggested the presence of intelligent life. Many scientists and politicians doubted this, calling it some sort of natural phenomenon, but the term "signal" stuck. Unmanned probes were sent to the planet, and they confirmed that a signal was produced during local daytime, but the source remained a mystery. To resolve things once and for all, the Space Council based on Luna City decided to mount a manned scientific expedition to Mercury, a strategy advocated by a news-hungry media and, curiously to some, by the Space Navy, which provided joint support. Now, in 2130, a crew of eight men and women was about to land on the planet, with the purpose of exploring the surface and resolving the mystery of the signal phenomenon.

"Touchdown!" Anthony Brown announced as the ship shook with a slight bump, then all became quiet as he shut down the engines.

"Nice job," Captain Kilborn said.

The pilot grinned, his dark moustache spreading across his thin, rugged face. Born in Station Alpha, the oldest of the Earth-orbiting satellite cities, the 49-year-old over-achiever never tired of hearing that his work was appreciated.

"Thank you, sir," he said.

The crewmembers activated the motors that swiveled their couches to accommodate to the vertical orientation of the ship. When this was accomplished, the small personal computers in front of them slid out of the way on their telescopic guiding arms, and the restraint harnesses automatically unbuckled. Kilborn looked at his crewmates.

"All right boys and girls, we finally made it. Let's do our landing check and get ready for an excursion after the Sun rises."

Tilda Chang stretched her arms upward, her lithe figure making her jumpsuit appear baggy. With a smooth complexion and short black hair arranged in a page boy cut, the 40-year-old engineer and communication officer looked half her age. She got up slowly from her acceleration couch.

"It's nice to be able to count on some gravity again," she sighed.

"Easy does it, Tilda," the crew physician said from the chair next to her.

She smiled at the tall Swede. With his blondish hair and blue eyes, she found him very attractive, and a fleeting thought crossed her mind about what he would be like in bed. *But his romantic interests are directed elsewhere,* she thought.

"Yes, all of you, listen to Dr. Ahlgren," Kilborn said. "After all these weeks in micro-*g*, let's not over-do it."

In recent years, speedy but expensive fusion-powered space ships had been developed to be used for missions to the distant outer planets and the Kuiper Belt. However, for missions to closer planets and their moons, it was more economical to use slower nuclear fission-powered vehicles. Consequently, for the Mercury expedition, it had taken the crew 4 months to make the trip, much of it coasting in micro-*g*.

"I can handle a little gravity," Chuck Morgan boasted. "It's only one-third that on Earth."

"Actually, 38 %," interrupted Akira Sato. The 33-year-old electronics expert from Berkeley liked to be precise.

"Still," Lars said, "Let's take it easy today."

"Tilda, radio mission control and let them know we landed successfully," Kilborn ordered.

As she activated the comm, Lars squeezed by his crewmates on his way to the elevator that led to the medical section down below. He needed to get things ready for the post-landing crew physicals. Anthony and the Captain checked the central console to verify that all of the essential systems were functioning. Chuck and Akira began their walk-through inspection of the ship. Boris Baganov, the mission geologist, glanced out the left view port and licked his lips in excitement at the upcoming visit to the surface of the planet.

Lost in all this activity was Samantha Evans. In fact, she had felt lost throughout the entire mission. When the crew for the expedition was being selected, someone in mission operations brought up the idea of including an astrobiologist. After all, primitive life forms had been found elsewhere in the Solar System, so perhaps this also would be the case for Mercury. Most scoffed at this idea. A geologist to study rocks and an electronics expert to interpret the mysterious electromagnetic signal were justified scientific crewmember positions, but Mercury was thought to be a dead planet. Nevertheless, no one knew for sure, so the position was added at the last minute. Samantha was an excellent astrobiologist. In addition, the crew selection psychological tests showed her to be an unusually empathic person, and it was felt that she could also assist the crew physician to manage emotional and interpersonal issues during the mission, so she was picked to fill the last crewmember slot.

She looked out the right view port, trying to make out some details. She saw little, however, since the Sun had not yet risen over the eastern mountains that rose between them and the rim of the basin. But she noticed the glow of the Sun's zodiacal light low on the horizon, signaling the impending dawn.

"Message sent to mission control," Tilda announced. "We should receive a reply from Luna City in about 16 min."

"Great," Kilborn said as he continued to monitor the readings. "Any sounds coming from the planet?"

All the crewmembers stopped their activities and waited.

"No, Captain, no signal sounds or anything that stands out," Tilda said.

"Set the receiver to the targeted EM frequency ranges, and put the system on automatic alarm. I want to know about anything that comes in from the surface."

"Yes, Sir," she said.

"It's dark out there," said Akira, pointing to the outside. "The radiation source has only been detected during daylight conditions."

"I know, but the Sun will be up soon, and we need to be vigilant."

Everyone returned to their landing duties, but all thoughts were on the mysterious signal and their reason for being on this hostile world.

Station Delta, 2126

The mood was festive in the Sector P party room. It was crammed with 145 people, the maximum allowed. They were gathered to celebrate Samantha's return from Europa, where for the past 2 years the young post-doc astrobiologist had been studying the native life forms living in the subsurface water. Her scientific paper correlating the genetic similarities between these organisms and the microbes on Mars had created quite a stir in her field, and she had returned home to take a well-deserved break.

Her father stood up, glancing at his daughter, who had grown up to be a fine-featured beauty. *Just like her mother,* he thought. He raised a glass of champagne and proposed a toast.

"To my daughter," he said, "who has just returned from Titan. We are happy to have her home!"

Samantha glanced at her mother, who showed no reaction to her father's obvious mistake in naming the wrong moon.

"She's been away for, let's see, 3 years, or is it 2. Anyway, Kalisha and I have really missed our only child. To Samantha!"

"Hear! Hear!" murmured the crowd to the sound of tinkling champagne glasses.

After the party ended, Samantha accompanied her parents back to their apartment, where her former sleeping area had again been converted to a bed space from its recent use as a small den. When her father excused himself to go to the bathroom, Samantha moved close to her mother. Samantha began to chew on her lower lip, a mannerism she displayed when she was distressed.

"How is Dad doing?" she asked.

Her mother sighed.

"The screening tests all came back positive for early-onset Class 4 Alzheimer disease, the kind we can't cure. The doctors can slow it down a bit, but the course is insidious."

She began to tear up, and Samantha put her arm around her mother.

"Mom, I'm sure the doctors will give Dad the best treatment. I just read a research paper from San Francisco that outlined another type of gene therapy for treating Class 3 Alzheimer's. They can stop the amyloid plaque production, and there are even cases of neurofibrillary tangle reversal. Hopefully, the same kind of therapy will be developed for Class 4."

"I hope so, dear. Dad's memory has really started to slip. He's still able to work, but his activities have been simplified and are closely supervised. There's a real race against time for a proper treatment, since his disease is progressing rapidly."

"We must be positive," Samantha stated as her father returned.

"And how's Daddy's little girl? All grown up, and famous. Tell us some more about your research on Titan."

"It was Europa, Dad, and it went very well. But I'm a bit bushed now after the great party you threw for me. Let's talk about my work in the morning."

"OK, sweetie," he said. "In the morning. Remind me if I forget to bring it up again—my memory isn't what it used to be."

"I know, Dad, and I'll remind you tomorrow."

She headed for her sleep area as her parents went off to their room. As she got ready for bed, Samantha couldn't help but feel sad, despite the festive reason for her being home. Once virile and sharp, her father had visibly declined in his mental faculties. He retained his sense of humor and optimism, but she knew that underneath the facade he was terrified of losing his memory completely and becoming dysfunctional.

Damn, she thought, *the one kind of Alzheimer's we can't treat.*

She slept fitfully that night.

Mercury, Days 2 and 3

To measure time on Mercury, the *Hermes* crewmembers continued using the 24-h Earth Standard Time that had been used on the ship during the transit from Luna City. During their second day in Caloris, they unpacked and organized their supplies and equipment to accommodate themselves for planetary living. Chuck and Tilda were tasked with assembling the two rovers, their means of venturing forth on the surface.

After the two engineers completed their work, the vehicles each resembled a large monster truck. Four huge wheels with deep treads were held in place by a strong white chassis that supported the air-tight passenger compartment. Highly efficient solar panels extended up from the flat roof. These soaked up enough solar rays to guarantee a continuous source of energy during the long Mercury day, when the Sun was continuously in the sky for 88 Earth days. *Rover 1* was the smaller of the two vehicles and was designed to carry up to six people on long excursions. It was equipped with bunks and enough storage space for food and water to last for up to 6 weeks. The larger *Rover 2* also had room for six people, but in addition it had an attached cargo compartment that provided enough space to carry equipment and heavy supplies on shorter trips.

****

On the morning of the crew's third day on Mercury, the Sun appeared just above the eastern mountains. The temperature was still way below zero as a result of the preceding cold nighttime temperatures, but it would rise as the

Sun transited the sky. The batteries on the rovers had received their initial charge on the ship, and the two vehicles were maneuvered down a ramp leading from the cargo hold to the surface.

Next, the crewmembers set up the rover tent adjacent to the *Hermes*. This was large enough to house the two vehicles outside. Made of a special alloy to tolerate the heat and radiation from the Sun, the tent could be opened at the top to allow the rover panels to receive light, or it could be closed, sealed tightly, and pressurized. A built-in oxygen generator and temperature regulation system allowed the crewmembers to perform routine maintenance activities or make repairs without a space suit.

Rover 2 was driven into the tent. *Rover 1* was left outside and supplied for the day's outing.

"The first excursion will involve me, Chuck, Lars, and everyone on the science team: Boris, Akira, and Samantha," Kilborn said. "Anthony will watch the ship, and Tilda will monitor the comm. Now that the Sun is up, we may receive a signal from whatever it is that we are here to investigate."

"Aye, aye, Sir," Tilda and Anthony said, almost in unison.

The six explorers went to the suiting area to put on their space suits. Samantha watched Kilborn as he arranged his gear. The 56-year-old Captain commanded authority, with his salt-and-pepper hair, dark green eyes, and authoritative features.

At his age, my father was starting to experience memory problems, she thought. *No one paid them much attention. "Too much work and not enough sleep," people said. Maybe if we had gotten him medical attention sooner. . .*

She turned to her own gear, stepping her trim figure into the space suit and tucking her long brown hair into her helmet as she checked the seals.

"Boris, unless Tilda reports a response from the signal source, this trip will be your show," Kilborn's voice echoed in her helmet intercom. "We'll collect some rocks and samples of the regolith for you to examine, and generally just explore the area near the ship. Let us know if you see anything interesting to pick up."

"Da, Captain," the Russian said in a muffled voice.

Suited up, they went outside and entered *Rover 1*. For the next 2 h, they drove a spiraling course outward from the *Hermes*, stopping here and there to pick up samples. Long shadows greeted them, slanting away from the low Sun. To the west, the surface was more level, extending to a horizon that seemed close and sharply delineated against a black, star-filled sky.

"How do things look, Boris?" asked the Captain.

"Like we expect. Surface shows lava deposits from old volcanic activity, and rocks probably iron-poor calcium-magnesium silicates. But I will analyze on *Hermes.*"

"Sounds good. Tilda, anything coming from the expected direction of the signal?"

"No, Captain," she responded from the ship. "No unusual frequencies detected from the west."

"Sorry, Akira. It looks like there's nothing for you to do today."

"I'm disappointed, but the daytime period is just starting," the physicist responded.

"You're right about that. OK people, let's head back. We had a nice first outing, and there will be many more chances to investigate the signal in the future."

Chuck turned the rover toward the ship. Samantha looked out her window toward the darkened western horizon, wondering what they would find out there and what the future held in store for them.

Station Delta, 2129

The holophone buzzed. Kalisha sat down in front of the holographic projector in the center of the kitchen table. The display on the projector indicated that it was her daughter calling from Luna City, so she activated it after sliding her gin and tonic to the side out of visual range. A hologram of her daughter's head and upper chest formed over the projector.

"Hi, Mom. I'm calling to wish you a Merry Christmas."

"Hello, dear. Thank you. How've you been?"

"Very busy. In fact, I have to leave for a meeting in a few minutes, but I wanted to call while I had the chance."

"Nice to hear from you. Is there any possibility you could come home for a visit before New Year's?"

Samantha chewed on her lower lip.

"I won't be making it home for the holidays. Mom, I have some news. I was selected to be one of the crewmembers for the Mercury expedition. I have to stay here and get ready for my training, which will start next week, on January 2nd."

"Congratulations, dear. See—I told you they would need a famous astrobiologist on board, just in case they find some little green men."

Samantha laughed.

"I doubt that that will happen. But after we investigate the EM source, there's a chance that we'll go up to the north polar region to take a look at the craters with ice in them. Perhaps there will be some microbes or spores embedded in the ice."

She paused.

"But there may be a break in my training later in the year before we launch. If so, I'll fly home for a few days."

"That would be wonderful, dear."

"How's Dad?"

"The doctors put him in the nursing facility. His memory is really bad now. You remember Sam Johnson, his former supervisor? He stopped by last week, and Dad didn't recognize him. Dad just can't take care of himself, and he's becoming too much for me to handle. But the nursing facility is close, in Sector R. He remains happy and comfortable, and the nursing care is very good."

"How're you doing, Mom?"

"OK, I guess," she said, eyeing her drink at the corner of the table. "After nearly 37 years, I feel like I've lost my partner in life. He's there physically, but psychologically he's somewhere else. It's not easy for me. I have my friends, and I keep active at work, but it sometimes gets lonely when I'm in the apartment alone."

"And I'm always away," Samantha added mournfully.

"Hush, dear, you have your life to lead. And I know you try to get home when you can. Anyway, we can always talk on the holophone, and I really enjoy the visits that you're able to make."

"It'll be hard to talk to you when I'm on Mercury, Mom. The two-way communication time delay can be more than 23 min when the two planets are far away, on opposite sides of the Sun. So we really can't chat in real time."

"I'm sure we'll find a way to communicate, dear."

"I guess so. Anyway, tell Dad hello from me when you next see him."

"I will. He always reacts with a twinkle in his eyes when I mention your name, although I'm never sure how much he remembers about you, his 'little girl.'"

They both smiled.

"Anyway," she continued, "I'm finding ways to cope."

"Mom, let me know if I can do anything. Perhaps we can contact a specialist on Earth, where they're doing all that research on the brain. What with all the heat and pollution affecting everyone, it's a good place to study brain functioning."

"I think the doctors here are fine," her mother said. "They seem to be up on the latest Alzheimer research, especially since so many of us here are expected to live past 100 years. What with all the gene therapy and new medications, and the surgical advances. . .imagine, partial brain transplants! It's too bad your father has a disorder that affects so much of his brain. But who knows what the future will bring."

"Yes, we can only hope."

Samantha glanced at the organic computer watch implanted in her wrist. "Well, I have to go now. I'll keep you informed of my schedule."

"Yes, do that, dear. And good luck with your training."

After her daughter disconnected, Kalisha looked around the apartment, which looked familiarly barren. Memories of better days came to her, when her family was together. Despite her hopeful talk, she doubted that her husband would improve or even stabilize in his downward spiraling course. She simply had to adjust to the situation.

Yup, she thought, *I just need to find ways to cope.*

She gulped down her drink, turned toward the foodbot, and ordered herself up another gin and tonic.

Mercury, Day 14

Over the next 10 days, the crewmembers undertook several more excursions on Mercury. One lasted 3 days and was aimed at testing *Rover 2*'s life support systems and giving Anthony and Tilda some experience working on the surface of the planet. The Sun continued to rise in the east, but there was no hint of a new EM radiation signal being produced in the west.

During their fourteenth day on Mercury, the crew gathered for dinner in the dining area. This was located in the mid-deck, along with the exercise equipment, the leisure area with its holographic projector, and the medical and geological laboratories. Since the *Hermes* was positioned vertically in its landing configuration, the ergonomically-constructed furniture and equipment had been adjusted to accommodate the new "down", which was toward the surface of the planet. Consequently, the fore-deck now was perceived to be above them. This area was occupied by the Command & Communications Center, which included the ship's various operational and life support control consoles, the outside view screens, and the crewmember acceleration couches. Below the mid-deck was the aft-deck, with the eight crewmember sleep pod rooms and the two bathrooms, and under this was located the main airlock and space suit stowage room. Beneath this crewed area were storage facilities and the giant nuclear fission engines that provided thrust for the ship.

All of the levels were connected by a central shuttle surrounded by a transparent protective tube. This moved crewmembers along the length of the ship in either its horizontal flight mode or vertical landing mode, where it functioned like an elevator. An emergency ladder was discreetly embedded within one of the walls for use in the event of an elevator failure. However, the crewmembers found the space around the ladder to be a rapid way of propelling themselves from level to level when the ship was in microgravity.

"So, when are we going to hear the signal?" Lars asked no one in particular while he microwaved a synthetic steak dinner. "The Sun has been up for 2 weeks."

"But the temperature is only 30 °C," Akira said, munching on his tofu sandwich. "Maybe whatever is producing the EM radiation needs more heat."

"In that case, it won't be long," Chuck said from across the table. "The temperature is going up fast outside. It will reach the boiling point of water in a week."

"I have a question," Samantha asked, sitting next to Chuck. "In our briefing on Luna City, we were told that the signal was first detected by a flyby probe and that it had a low frequency. How did it happen that the probe was able to detect it in the first place?"

Akira smiled.

"Actually, that was a main part of its mission. We've known for some time that 'extremely low frequency' EM waves below 100 Hz are produced on Earth, such as during earthquakes, volcanoes, and lightning strikes. Early in the last century, these waves were also detected on other planets and moons, like on Titan and Io. We can learn a lot about a heavenly body from its ELF signature. For that reason, the Kuiper Belt probe launched in 2127 included special equipment to see if these waves could be found in the bodies located there. The detection of ELF waves transmitted from Caloris was an unexpected occurrence."

"And no one knows for sure what the source of these waves is, correct?" Samantha continued.

"That's right," the electronics expert responded. "The Space Navy has sent probes over the suspected area, and although they've detected the signal, there have been no reports of a visual sighting of anything unusual on the surface. So. . ."

"That's why we're here," Samantha added.

She looked over at Boris. "Any ideas about what's causing the signal?"

"Nyet," the geologist said. "But whatever it is, probably some sort of local process maybe is involved. It could be on surface, or maybe underground. ELF waves are good to penetrate solid matter and ocean water—U.S. and Russia once used these to talk with submarines. But to make enough power to detect from space. . .there must be a lot of energy. Since ELF waves only are produced on Mercury during daytime, maybe energy is solar. Maybe underground volcanic vent is involved. Who knows?"

He paused, taking a sip of coffee.

"We must wait for surface to heat up and. . .how you say. . .soak up the rays!"

Everyone laughed.

During this discussion, Samantha had been glancing at Lars to see if he might have something to contribute to the discussion. During the outbound journey, she had formed a special friendship with the 38-year old physician. He too had been a late selection for the mission, in part due to the death of his wife in an air car collision on Earth 8 months before launch. But he had been psychologically cleared and was picked for the mission due to his experience in dealing with space crews exposed to the intense radiation around Io that was generated by Jupiter. She felt she knew his moods, and he looked worried as he sat down with his cooked steak dinner in hand.

"Lars, you know something about radiation," she asked. "What do you think is causing this signal?"

"I don't have a clue about what's causing it, but I have some concerns about our safety. Unlike high frequency ionizing EM radiation, which breaks up molecular bonds and causes tissue damage, low frequency waves can produce thermal effects that can fry cells if the radiation is intense enough. They can also produce non-thermal effects, like tingling or pain in the skin and flashes of light in the eyes. At the kinds of intensity that Akira and Boris are talking about, we need to be careful about being exposed. The ELF waves could penetrate our space suits, or even the walls of the rovers, or *Hermes*, or maybe pass through intervening hills..."

"Safety is paramount," Kiborn interrupted after just walking into the dining area. "We landed some distance away from the likely source of the signal. If it starts up again, we'll monitor it carefully for intensity and direction. If it increases, and things get dicey, we'll abort the mission and take off for home. I don't plan on adding my name to the list of captains who have lost crewmembers."

There was a momentary silence as everyone took this in. Samantha noted that Anthony looked especially thoughtful. He was looking down at his meal, averting direct eye contact with his crewmates. Chuck then spoke.

"Sir, have there been deaths from this signal?"

Kilborn scowled, then looked directly at him.

"New expeditions can be dangerous. Plenty of them have resulted in injury or death. I don't plan on that happening to us."

He turned towards Tilda.

"I want you to auto-monitor Caloris continuously for signs of unusual EM radiation and to instruct the computer to alert you every time an anomalous signal is detected. I want to be informed immediately, night or day."

"Aye, aye, Sir," she said.

"We can't be too careful," he concluded.

He turned his back to the crew and went over to the foodbot to select something to eat. The discussion had ended. Anthony stood up, took his plate of half-eaten food to the robowash, and walked over to the central elevator to go up to the C & C Center. The rest of the crewmembers exchanged puzzled glances at the sudden ending of the discussion, but no one spoke further of the signal as they ate in silence.

Mercury, Day 15

The next morning, Samantha woke up early from a fitful sleep and decided to go up to the medical lab. Besides assisting Lars with his medical and psychological evaluations, she had been trained on obtaining and analyzing blood samples taken from the crewmembers. As she exited the elevator to the mid-deck, she was surprised to see Lars in the lab.

"You're up and about early," she said.

"I decided to review the blood chemistry reports over the past 2 weeks to see if there were any kinds of changes due to radiation exposure. I know that Tilda is monitoring for signals, but I thought I would just double-check. I didn't want to alarm anyone, so I planned to conduct my review before people came up for breakfast."

"If you can wait a second, I'll run the latest blood samples through the analyzer so that you can add these results to your figures."

She went over to the freezer to get the samples.

"I didn't sleep too well last night," she said as she returned. "I don't know what to make of what the Captain said about crewmember injuries and deaths, or about captains who lost crewmembers."

"Yeah, I know, it puzzled me too. I'm sure he was speaking in generalities. But whatever the reference, the two of us need to be vigilant. We're responsible for the health of the crew."

Samantha placed the samples in the analyzer to be scanned. She turned to Lars, who had been watching her as she moved around. She blushed a little, then said:

"I guess I'm surprised that neither Akira nor Boris know more about what could be causing the signal. Akira is a real whiz-kid who has degrees in electronics and physics, and Boris has been to Mars and several asteroids and planetary moons studying their geology, yet the source of our EM signal seems to have eluded both of them."

"I agree," he responded, "but this whole mission seems to have an air of mystery to it. Our briefings were very vague about the signal, and the theoretical discussions of ELF radiation seemed pat to me and more relevant to Earth than to Mercury. I wonder if we're being told everything."

"Yes, me too."

Samantha paused for a moment to record the readings from the blood scans, then continued.

"At any rate, the mission has gone smoothly so far."

"Indeed. So far, so good."

"I'm feeling a little homesick," she stated. "Mercury is such a foreign place. How about you?"

He looked a bit wistful.

"You know, after my wife died, I wasn't sure about leaving Earth so soon. But I was based at the Astronaut and Space Services Center in Luna City, and I had to return to my duties there. The Chief Flight Surgeon alerted me to the fact that I may be selected for this expedition, so I quickly became involved. This actually helped me in getting my mind away from Laura."

"You were married a long time, yes?"

"Nearly 14 years. We met in medical school on Mars City, where both of us were born. We decided to get a marriage contract to improve the chance that we would be assigned together during specialty training. It worked. We both were selected to go to Houston to be trained as flight surgeons."

"What was it like for you there?"

"It was hard to adjust, what with Earth's high gravity and the heat and pollution, but it worked out. Laura sub-specialized in Psychiatry, and like my father I went into Radiology in order to study the effects of radiation in space on the human body. Afterward, we were stationed at the A.S.S.C."

He paused, a wistful expression on his face, then continued.

"Laura was on vacation visiting some friends in New York City when a drunken air car driver slammed into her vehicle. The fool was joy-riding under manual control and apparently had disengaged the crash-proof system. He and Laura both died instantly."

Samantha was touched. This was the first time Lars had spoken so intimately of his wife's death. She went over and gave him a hug.

"I'm really sorry about what happened."

Surprised, he leaned back and said: "Yeah, thanks, but at least we had no kids. We were so busy with our jobs that we put it off. So there was no one to miss their mother."

"Just you, missing your wife," she said as they separated.

"Even now, a bit, I guess," he responded. "What about you—do you miss not having a husband and children?"

"A little. But I had my career, and I enjoyed studying the microbial life on Mars and Europa during my fellowships. I needed to travel a lot for these activities. I just didn't feel comfortable committing to someone or having children alone or through a surrogate, so I postponed the whole marriage and children bit. And then my father really got sick—he has the incurable type of

Alzheimer's. Being an only child, I felt that I needed to support my mother in dealing with my father as he declined. Although I was not home much, I called frequently and visited as often as possible. Mom seemed to enjoy this contact. The communication time delay on this expedition has complicated things, but we've learned to compensate and get the message across."

She chewed on her lower lip as she stared off for a moment, then sighed and continued.

"But in a way, the expedition has been good for me. It's gotten my mind off of things at home and allowed me to focus on visiting a strange but interesting planet."

"And for me," he said smiling as he took her hand. "I've really enjoyed getting to know you more as well."

"The same for me," she responded. Their conversation was interrupted by the sound of the elevator bringing the other crewmembers up for breakfast.

Station Delta, 2130

Samantha walked briskly down the broad Main Corridor. Circling inside the huge ring of the station, it connected all of the sectors. She passed a number of apartments in Sector R, which were located against the outer wall to take best advantage of the centrifugal force that provided the nearly 1-g in the station. She entered one of the business areas and looked for her father's nursing facility. She only had 2 h before the shuttle left to take her back to the A.S.S.C. for further training. She was grateful for even this brief amount of time, since her departure for Mercury was to take place in 10 days.

Spotting the facility, she entered and asked the desk nurse to take her to her father. She was escorted into one of the adjacent rooms. Her father was sitting up in bed, staring at the hologram being projected in front of the wall, but likely not following the plot of the soap opera that was being shown. He was pale and thin, his grayish-blond hair disheveled, and he was dressed in a blue pajama set. Samantha noticed a cord attached to his belt and to the side rails, which prevented him from wandering too far away from the bed.

"Hello, Dad," she said.

He looked at her vacantly, as if she were a stranger. Then something clicked, and he smiled.

"Hello darling, how are you?"

"Just fine, Dad. How are you?"

"I'm doing just great."

Noticing the untouched luncheon tray next to him, she asked: "How has your appetite been?"

"Just fine. The food is good here. How are you?"

"I've been very busy. We leave for the planet Mercury next week, and I wanted to see you before I left. Mom said that you would be in your room at this time."

"Yes, I'm here. You're leaving for Mercury? When do you leave?"

"Next week, Dad. We'll be gone for nearly a year."

"That's nice."

He looked over at the food tray.

"The food is good here. I like the food."

"I'm glad, Dad. How are you feeling?"

"Terrific. Everything is great here. I like the food."

"Is there anything you need?"

"No, I'm just fine. You're leaving for Mars? When?"

"Mercury, Dad. I leave next week."

"Oh. How long will you be gone?"

"About a year. Mother will be coming in to see you, however."

"That's nice. And how is your mother?"

"She's fine. You wife Kalisha is fine. She says that you're getting good care here."

"Yes, I like it here. I'm glad Kalisha is fine. Will she visit me?"

"Yes, she's been coming in regularly, and she'll be visiting you in the future as well."

"Good. I hope your mother comes to see me, too."

Samantha felt the tears well up. She took a napkin from the food tray and dabbed at her eyes.

"What's wrong, darling? Are you sad?"

"Nothing, Dad. My eyes have been burning lately—maybe I'm a bit allergic to something."

"Oh, I'm sorry to hear that something here is bothering you. Where do you live?"

"I don't live here anymore. I'm based in Luna City, where I'm being trained for the Mercury trip."

"You're going to Mercury? The planet Mercury? It's hot there, isn't it? Where are you staying?"

"Yes, it will be hot. My crewmates and I will be on the first expedition there, and we plan to stay in the ship that transports us to the planet's surface."

"That's nice. How long will you be there?"

"Not too long. . .Dad, let me help you with your lunch."

She moved the food tray to his bed and helped him eat. He ate sparsely, choking every so often even though the food had been pureed. When he finished, he started looking at the hologram again. Samantha decided to just sit quietly next to him. When the show ended, she checked her computer watch and decided that she had to leave in order to catch her shuttle to Luna City.

"I have to go now," she said.

"Oh, do you have to see some other residents? Thank you for your help. I imagine you are very busy working here."

"Dad, remember that I'm your daughter, Samantha. I just came in to this nursing facility to visit with you a bit. I'm leaving for the planet Mercury in 10 days, so I won't see you for another year or so. My mother—your wife Kalisha—will be in to see you, however."

"Great. But I'm sorry you won't be here. How long will you be gone?"

"About a year. But I won't forget you. I'll try to send you some messages through Mom that she can read to you while I'm gone."

"That would be nice. I always like to hear from Daddy's little girl!"

Stunned at this comment, she quickly regained her composure, smiled, and gave her father a kiss. She waved goodbye. As she left the room, she glanced back to see her father waving and smiling. Fighting back the tears, Samantha hurried out of the facility and toward the docking area to catch the shuttle.

Mercury, Day 21

Day 21 was a milestone, with the outside temperature reaching 100 °C, the boiling point of water (if there had been water on the scorching surface of Caloris). But still no sign of the mysterious signal.

Chuck, Samantha, Akira, Anthony, Boris, and Lars left for an excursion in *Rover 1* just after breakfast. Their route took them due west, toward what they thought might be the source of the signal based on earlier flyby data. The landscape was barren and strewn with silicate rocks and craterlets. The Sun was 45° above the eastern horizon. They were 14 km from the *Hermes*.

Sitting in the front seat next to Chuck, Lars glanced at the radar screen and then looked up.

"Chuck, that's an odd image."

"Where?" Chuck said, glancing down. "Oh, I see it."

He pointed to an elongated blip on the screen.

Suddenly interested, Boris said: "Maybe we can go there to have look."

"OK," Chuck answered, "we're on our way."

They headed toward the northwest for 200 m.

"There it is, shining in the sunlight," Lars said, pointing slightly to the right.

They drove over to the object, and Chuck stopped the rover. Before them was a metallic, deep blue, rod-like object nearly 3 m long that was connected to a wider, darker piece of material that had a jagged edge.

"That looks man-made," Samantha said.

"It sure does," Chuck answered.

The six of them put on their helmets, exited the rover, and walked over to the object.

"It looks like an antenna of some kind connected to a piece of solar panel," Chuck said. "What's that doing here?"

He looked at Anthony.

"Could this have come from some kind of space vehicle?" Chuck asked.

The pilot lifted up the pointed end and examined the object with great interest.

"Hmm, yes, maybe."

"Boris, what's it made of?" Lars asked.

Moving closer to it, the geologist responded.

"Some kind of metallic alloy. And for sure not natural to area."

"Did it come from Earth?" Chuck asked.

"There, or from place that makes such things, perhaps on Moon or Mars."

"By its color scheme and shape, I suspect it was broken off from a Space Services vehicle," Akira intervened.

"Military…if so, what's it doing here?" Chuck asked, looking at Anthony.

"I don't know."

"What's your best guess? You're an officer of the Space Services."

"Maybe from a civilian probe of some kind."

"We're a long way from the normal commercial flight paths," Akira said. "And the deep blue color is characteristic of ships in the Space Navy, isn't it?"

"I said I don't know!" Anthony blurted out, with wide eyes. "I don't even want to speculate. You should speak with the Captain."

"But what will he know about it?" Chuck asked.

"I don't know…talk to him," Anthony repeated, his voice rising. He continued to look at the object, not making any eye contact with Chuck or anyone else.

He's hiding something, Samantha thought. *But what?*

"All right, suit yourself," Chuck said. "Let's put this object in the rover and report back to the ship."

****

After a late lunch on the *Hermes*, the crewmembers all gathered before the mysterious object, which had been placed on the examination table.

"So, what do you think it is, Captain?" Chuck asked.

"I would say that it's a piece of a flyby probe," he responded. "Probably one of the vehicles sent by the Space Council to investigate the signal."

Anthony glanced over at him, but Kilborn averted his eyes and didn't say anything further.

"I don't recall hearing anything about a Space Council probe being damaged," Chuck said.

"Plus, its blue color is characteristic of a Space Navy vehicle, right?" Akira added. "That would put it in the realm of a Space Services mission."

"That's true, but there's no official report of a Space Services-sponsored vehicle crashing in this area."

"Like Anthony, you're an officer of the Space Services, so I guess you would know," Lars said.

"Yes," he responded. "Actually, any one of several groups could have sent a probe here: the Space Council's Science Team, a political independence group from Mars or a gas giant moon, a commercial mining company. . .anyone."

There was silence.

No one wants to push the Captain, Samantha thought.

"Anyway, let's put this thing in storage for the technicians to look at when we return home," Kilborn said.

There were no objections, and the group went on to discuss other matters.

Luna City, Day 21

The Space Services Conference Room was buzzing. The variety of uniform colors indicted that all of the military groups were represented. No civilians or representatives from the Space Council were in attendance. The meeting was convened by Admiral Harvey of the Space Navy. With his gray eyes, white hair, and heavily starched blue uniform that displayed numerous colored battle ribbons, he conveyed an aura of authority that befitted his rank as Chief of Staff.

"Gentlemen and ladies, please give me your attention. Our emergency meeting is hereby in session."

Everyone became silent. Harvey continued.

"An hour ago we received a coded message from Captain Kilborn on board the *Hermes.* He said that during a routine exploration mission the crew found a piece of the *Endeavor* on the surface of Mercury. He believes it to be a blue antenna attached to a solar panel. Suspicions have been raised by some of the crewmembers that it came from a Space Navy ship."

"How could that be?" asked General Suzuki of the Army. "We were told that the *Endeavor* was destroyed more than a hundred kilometers from the *Hermes* landing site."

"It was, but apparently parts of it traveled far and wide, and one piece happened to be discovered."

"What do we do now?" asked Major General Santini of the Marines.

"We must maintain complete secrecy, or panic could ensue, not to mention our credibility. We're facing a potential threat to our peace and stability, and it may escalate in an unpredictable manner."

"Yes, indeed," added Rear Admiral Gazenko. "It's just this sort of lack of confidence that will encourage the Mars Secession terrorists to accelerate their attempts at planetary independence."

"Maybe they were responsible for what happened to the *Endeavor*," Santini said.

Harvey shook his head. "I don't think that even they are sophisticated enough to successfully do such a thing. No, it was a local action that likely was related to the signal from the surface of Mercury. We need to fully understand what we're dealing with."

"What do we tell Kilborn?" asked Suzuki.

"I already advised him to stay the course," Harvey said. "He has his orders to explore and attempt to understand the nature of the signal. He is not to mention the *Endeavor* to any of the civilians in the crew. Right now, only he and the pilot, Lieutenant Commander Brown, know about the *Endeavor* incident. Even the Communications Officer, Major Chang, has been kept in the dark. She is only following orders on a need to know basis."

"And if one of the civilians finds out?" asked Santini.

"Well, General, we might have to declare war on the *Hermes* and invade Mercury with your Marines!"

Everyone laughed.

"What if the *Hermes* itself becomes a target?" asked Suzuki.

"That's tricky," Harvey responded. "Unfortunately, this mission has been well-publicized and pictured as some sort of mystery involving a strange signal. We have planted the story that it likely represents a natural phenomenon occurring on Mercury, without danger to anyone. For that reason, we asked the Space Council to involve the Astronaut Division and to classify the mission as a civilian expedition, complete with top scientists. The military ranks of the senior officers on board were deemphasized, although the crewmembers know that Kilborn and Brown are in the Space Navy, and Chang is in the Army. Captain Kilborn has been told that if he perceives any danger, he is to evacuate the planet immediately and avoid any conflict."

"What if that scenario doesn't work and the *Hermes* and her crew become threatened in some way?" asked Santini.

"Then we'll have to consider a military solution. But for now, I suggest we wait and see what happens. Are there any further questions?"

Receiving no additional responses, Harvey adjourned the meeting.

Mercury, Day 30

During the next 9 days, the *Hermes* crew was kept busy by routine maintenance activities and three rover expeditions on Mercury's surface. More silicate rocks were collected, with Boris failing to find anything new in his analyses. During one excursion, Chuck and Akira tested a new portable X-ray machine that allowed them to image the ground several meters below the surface. Nothing special was revealed except for more rocks buried in the regolith. Tilda remained closer to the *Hermes*, testing a new kind of communications transmitter on behalf of the Army.

On Day 30, the temperature passed 200°. Still no signal.

"You know," Chuck said during their lunch break, "maybe we aren't going far enough looking for the source of the signal. We should extend our explorations further out."

"But how far?" Lars asked. "We know the signal should come to us from the west, but we don't know for sure if it's due west, northwest, or southwest, and how far away its source may be."

"Yeah, but we've been here a month now, and we have nothing to show for it but a few rocks."

"Well, we found antenna and solar panel," Boris said, smiling nervously. No one responded to his comment.

"Akira, any ideas about where we might find the source?" Chuck asked.

The electronics expert looked up from his bowl of algae soup.

"Not really. We need to receive a signal of some kind that can be picked up by the satellites we launched into planetary orbit before we landed. They will allow us to triangulate the direction of any ELF waves that are transmitted. But without a signal, there's not much we can do."

"Maybe we could do a fly-over?" Chuck responded.

"What do you mean?" questioned Anthony.

"Well, the signal was originally detected by a probe that flew by Mercury. And since the original episode, a few other probes have also detected it. Maybe we could take off in the *Hermes* and cross back and forth over Caloris hoping to pick something up."

"Well...," the pilot said, "with our nuclear reactor we don't have to worry about using up fuel, but there's no guarantee that a flyby will trigger anything from whatever it was that caused the signal in the past. Plus, the temperature continues to increase as the Sun rises, and we suspect that the signal's appearance is related to the surface heat in some way. Time is on our side. I think we're better off just waiting for the temperature to rise some more."

"It's already twice the boiling point of water," Lars said. "That seems hot enough to me!"

"But it can go up to over 400° here in Caloris," Anthony said, "so we have a ways to go until we achieve maximum temperature."

"What do you think, Captain?" Samantha asked.

"I agree with Anthony. Our orders were to land east of the presumed source just before sunrise, when the Basin was still cool from its months of exposure to local nighttime conditions. Then, we were to make exploratory excursions in the area by rover when the Sun came up and wait until the signal was detected."

"This is scientific mission," Boris said. "We are collecting and analyzing rocks, and Akira has done radiation measurement of surface. So, mission is fulfilling many science goals."

"And we know that the electronics in the rovers' walls and the ship's hull can shield us from the inhospitable conditions on the outside by deflecting the heat from the Sun and neutralizing its radiation," Chuck offered, "at least until now. That's good technological information for further expeditions to this planet."

"Samantha hasn't had much to do yet," Lars said. "You haven't found any life on Mercury, I take it?"

"Just questionable life. . .on the *Hermes*," she joked.

Everyone laughed.

"But seriously," Lars continued. "Any little microbes related to the rock samples?"

"Nothing," she said. "With all the heat and radiation on the surface, it's hard to imagine that any carbon-based life can exist here in Caloris."

"What about the crater ice at the poles?"

"Maybe. On Europa, there are multicellular organisms, but they're all found in the subsurface water, protected from the radiation, cold, and near-vacuum of space by the icy surface. We haven't detected any liquid water on Mercury yet. But perhaps we'll be able to explore the poles later on?"

She turned to Kilborn.

"What do you think about such a trip, Captain?"

"Maybe in time. But right now, my orders are to explore Caloris and to find the source of the signal. That's our top priority."

"Why is that given so much priority?" Lars wondered aloud. "I too would be interested in going up to the north polar region to take a look at the ice. It could be very productive, scientifically."

"And geologically," added Boris.

"I'm not going to question the orders," the Captain responded. "Let's all just do our jobs and get our primary mission goal accomplished, safely."

"You've mentioned the need for safety before, Captain," Akira asked. "Is there some sort of danger here?"

"All space missions are dangerous," he responded. He stood up and took his plate over to the robowash. The discussion was over.

Mercury, Day 33

To: k.evans@stationdelta.org (RESTRICTED USE)
From: samantha.evans@Mercuryexplore.assc
Subject: Hello again
Date: Mercuryexplore, Day 33

Hi Mom,

I haven't contacted you in several days because I'm lazy, not because we've been busy here. In fact, nothing much has been going on. The outside temperature is well over 200 °C now, but we still haven't detected our mysterious signal. The Captain has us making daily excursions in one of the rovers, and we have continued to collect rocks and regolith samples, although nothing new has been found. There hasn't been much for me to do. I've been helping Dr. Ahlgren with his crew medical evaluations, but even these are pretty routine these days, now that we're settled on the surface.

Lars—Dr. Ahlgren—is a very nice person, and we have become good friends. In fact, most of my crewmates have been very nice. The Captain continues to be authoritative, and the pilot seems to be holding on to some secret, but everyone else is upfront and sociable. Yesterday, we celebrated our geologist's birthday, complete with a terrible tasting cake created by our foodbot. It seems that the machine is pretty good at whipping up simulated steak and potatoes, but not so good at desserts. What a sad state of affairs!

How are you feeling? Are you over your cold, yet? You mentioned before that you were starting anti-viral medication. How are the algae farms doing? We sure could use a nutritionist here on Mercury to spice up the food a bit. But I guess this is a complaint common to all spacefaring crews.

How has Dad been? I presume his memory continues to deteriorate. Has he been his usual pleasant self? Sometimes people with his condition get irritable and explosive, but from your last message it sounds like Dad has continued to be even-tempered. You've indicated before that the staff members at the nursing facility are giving him good care. Are there any changes in his treatment regimen?

How have you been dealing with Dad's situation? I know that it's difficult for you, and I'm really sorry that I can't be there to help.

Well, that's all for now. I look forward to hearing back from you.

Love,

Sam

****

To: samantha.evans@Mercuryexplore.assc (RESTRICTED USE)
From: k.evans@stationdelta.org
Subject: Hello back
Date: Mercuryexplore, Day 33

Dearest Samantha,

Lovely to hear from you. It sounds like things have become pretty routine and that not much has been happening on Mercury. I'm sure that this will change. From what you've told me, the signal becomes active during elevated temperatures, and you are still far from the daytime high in Caloris. It seems strange to me to speak of 200° as not being hot enough, but I guess it's just a matter of perspective.

I'm glad you're enjoying most of your crewmates. Sometimes when confined people get bored, they start to get snappy with each other. This happens on Station Delta when we're restricted to our sector for construction or an emergency. But friends and loved ones are important. I'm especially glad that you've found Dr. Ahlgren to be a special friend. Isn't he the one whose wife died? He must feel lonely at times. It's hard to lose a spouse.

I feel that way sometime about your father. He's physically here, but the man I married and loved has disappeared. In a way, he's dead to me. At least he's not suffering. His physical health seems OK, and they try to keep him active. I only wish he would eat more. He gradually has been losing weight, which has some of the nursing facility staff worried. I've been worried as well. I have several friends who have been supportive, especially John Gomez, a widower who works with me in the algae farms. We go to dinner and parties together, and this takes my mind off of things.

There is one new development. They've started a music therapy program for your father and a group of other residents in the nursing facility who can move about. The music is accompanied by dancing and lights that change color in time to the music. I've attended some of these sessions. Your father seems to respond energetically to the different tunes they play, especially the happy ones. He also seems to enjoy the changes in the lighting. He smiles and moves to the beat and sometimes seems like his old self. On occasion, we've danced together. When the sessions end, he retreats again into himself. I know that he's alert and aware, and that he has the potential to respond to his environment. It's just that his memory is not working very well, although sometimes he asks about his "little girl". I tell him where you are and what you're doing, but it doesn't seem to stick. Nevertheless, he's comfortable and seems to be without pain or distress. Sometimes I feel that his disease is harder for me and my psyche than it is for him.

My cold has gone—the new anti-virals are really effective. My job continues to keep me busy. I feel very much appreciated when people tell me how good the food tastes. I never liked the taste of algae, but they are so much a part of our diet here that making the algae products palatable is an important undertaking. The hydroponic system is functioning well, and the biochemists are always coming up with some new genetic modifications to expand our food source. I wish I could suggest something to improve your situation on *Hermes*, but at least your stay on Mercury is time-limited, and you will be back to civilization and better food in no time.

You might have heard that the Mars Secessionists have blown up a terraforming facility just outside of Mars City. I guess they feel that as Mars become more comfortable for human life, the offloading of people from Earth will escalate. They seem to fear that Mars will become another polluted Earth. I just don't understand this reasoning. With proper birth control, and with the success of our fusion reactors in producing clean energy, this could be prevented, and Mars has a lot of land for settlement. Earth has really become an uncomfortable place on which to live, and I think humanity has learned its lesson. But I guess people just don't want to help others less fortunate than they are. Humbug!

I guess I've ranted on long enough. Let me know how things are going from time to time. Mercury sounds like a very interesting, if not hospitable, place. Keep in touch.

Love you.

Mom

Mercury, Day 34

The next day, Samantha reflected on the messages between herself and her mother. All personal "restricted" contacts between crewmembers and their families were completely private, even from the Captain and mission control. But she felt the need to talk about her feelings with someone else, so she sought out Lars after lunch. He was in the medical lab going over some tests. Since it wasn't yet time for the afternoon exercise period, they were alone on the mid-deck.

"Lars, can I speak with you a moment?"

"Sure, Sam," he responded, putting down his printouts. "Have a seat." He motioned to the two lab stools behind the microscope stations.

They sat down. Lars smiled at her.

"What's up?" he asked.

"Well, I received a message from my mother yesterday, and I wanted to talk with you about something."

"Is it your father? How's he doing?"

"About the same, although he's undergoing music therapy and seems to be enjoying it. But that's not what I wanted to talk to you about. My mother mentioned that there has been increased activity from the Mars Secessionists. They apparently blew up a terraforming facility near Mars City, where you grew up. I know about them from the news reports we've been receiving from Luna City, but I wanted to get your opinion of the situation."

"Nothing malicious they do surprises me." Lars said somberly. "This group really wants Mars to be independent, to secede from the Solar System States. They view terraforming activities as just another way to keep Mars in the fold, to allow it to become overpopulated by all the immigrants from Earth."

"But wouldn't the Martian citizens benefit from having a more Earth-like environment?"

"Certainly, but this terrorist group feels that the political cost is too high. In their view, the result would be the loss of planetary autonomy. Plus, they fear that a Martian version of the "Great Pollution" could someday occur. It's unlikely, given our experiences with what happened on Earth, and what to avoid in the future. But a significant minority of the local populace on Mars supports the terrorist point of view, if not their tactics."

"How organized are they?"

"Very. They have cells on Earth and on some of the asteroids and planetary moons. The rumor is that several Martian senators secretly support them as well. Although most of their terrorist activities have taken place on Mars, this group is capable of causing mischief anywhere in the Solar System."

"Even on Mercury?"

"What?" he asked incredulously. "I don't understand your meaning?"

"After we found that antenna, the Captain brought up the possibility that it could have come from a piece of equipment used by a Mars independence group. At the time, we were discussing some sort of flyby probe. But maybe the equipment is related to the strange signal we're here to investigate. Maybe the Mars Secessionists are somehow involved. Maybe they've built a machine to generate the ELF waves. Maybe..."

"Whoa," Lars said. "There are a lot of 'maybes' here. I can understand building a probe—this group has been able to manufacture small chemical-powered space ships—but so far as anyone knows, we are the first people to land on Mercury. I can't see them constructing a powerful ELF transmitter on the surface. Plus, even if they could, what would be the reason for it? The Secessionists are worried about Mars independence, not with what happens on a hot, desolate planet close to the Sun."

"I suppose you're right," Samantha said thoughtfully. "I guess I'm just feeling a bit paranoid. I keep thinking that we're not being told everything by the Captain, and it's made me begin to question a lot of things about this mission."

"I have some discomfort about this mission as well," Lars said, "but I don't think the cause for our concern is the Mars Secessionist group. Something else is going on."

They looked at each other. Lars reached over and gave her hand a squeeze. She smiled.

"Thanks for hearing me out," Samantha said. "I hope this doesn't result in a psychiatric alert being placed in my records on the basis of paranoia and faulty reasoning!"

They both laughed. Lars glanced around, making sure that they still were alone. Then he leaned over and gave her a hug. She became titillated at the warmth of his body. She impulsively kissed him on the lips.

"Oh my God, I'm sorry," she said. "I don't know why I did that."

She felt herself blushing, but he just smiled and kissed her back.

"I've been wanting for us to do that for some time," he said, "but I guess I wasn't sure that you would receive it positively. You're a special person to me, Sam, and I'm glad you broke the ice."

Just then, they heard the elevator, likely bringing someone to the mid-deck to perform their scheduled afternoon exercises. Lars winked at Samantha, and they separated. It was Tilda, dressed in her exercise clothes. She smiled at Lars but glared at Samantha, then she headed for the treadmill.

I guess she isn't too happy that Lars and I are here together, Samantha thought. She then went over to the refrigerator to pull some blood samples for analysis.

Mercury, Day 41

On Day 41, the outside temperature approached 300°. Just before lunch, Tilda went up to the fore-deck to check her equipment in response to the buzz of a comm alert message. Her eyes widened as she listened to her earphones, and she became excited.

"Captain!" she announced on the intercom, "I'm picking up something that's being relayed via the orbiting satellites."

"What is it?" answered the breathless voice of the Captain. It came from the comm link that originated next to the treadmill, where he was running a simulated quarter marathon. It was 11:42 ship's time. The other crewmembers were finishing up projects or resting in their sleep pods before lunch.

"A steady radiation source," Tilda responded. "According to the record, it started a few minutes ago as a faint EM pulse, but now it has intensified. It's not nearly as strong as the signal that was picked up by the flyby probes, but it's detectable. Triangulating it with the satellites puts it around 105 km away at 8.34° west northwest.

"All right! Finally...something," Kilborn said.

He activated the master intercom system. "Attention all crewmembers. We've picked up a signal from the west. Everyone...finish your projects or whatever you're doing and come to the mid-deck. Let's grab a bite of lunch, then we'll talk about going to have a look at what's causing the signal."
****

The signal steadily persisted during their lunch period. The Captain decided that all of them except for Tilda and Anthony would take *Rover 1* and head out toward the likely source of the radiation. They all felt great excitement and anticipation as they made their preparations.

At 1330 h, the six designated crewmembers left the suiting area and boarded the rover. As they drove toward the source of the signal, the blistering hot, dusty course toward the west remained flat, although at times Chuck had to swerve to avoid a silicate rock or drive around a deep crevasse. The vehicle was surprisingly maneuverable given its four large wheels with their deep treads, and they made good time over the surface of Caloris.

"The sensors indicate that we're getting close," Chuck reported after 2 h in transit. "The source is up ahead, at the summit of that mesa-like formation."

They reached their goal and circled its base until they found a relatively smooth and gradual pathway spiraling up to the top, which was some 90 m above the flat basin floor. Chuck stopped the rover, and the rest of the crewmembers put their helmets on and got out to explore while he remained inside to monitor communications from Tilda and keep an eye on the rover.

The exploration party slowly ascended, being careful with their steps. But the climb was easy in the light gravity. When they reached the top, they were on a wide plateau. Roughly in the middle was a lumpy, glassy circular mass about 50 m in diameter.

"That's the source," Akira said, checking his portable radiation detector.

"It's hard to see clearly in bright sunlight," replied Boris. "It looks like crystal...I can almost see through it. Maybe that's why it was not imaged by flybys."

"That, plus the size is just below the resolution of their detectors," Akira added.

Kilborn looked at him. "How about the radiation?"

Checking his portable radiation detector, Akira responded.

"There are waves of several different frequencies, Captain, but ELF waves are prominent. Fortunately, they're not intense enough to penetrate our suits."

"Then let's go take a closer look," Kilborn said, walking toward the mass.

As she followed the team closer to the source, Samantha could see that the mass was made up of sharply pointed crystal rods that reminded her of pictures she had seen as a girl of old-fashioned telephone poles on Earth. Each tilted vertically in a random direction and extended about 2 m out of the ground.

"It looks like a clump of giant rock candy," Lars said, laughing to break the tension.

In Mercury's low gravity, it only took a few steps for them to bound over to the nearest rod. Small shiny fibers of varying sizes connected the crystals, each one to several others. Boris removed the pick hammer from the tool belt of his space suit and chipped away some fragments, which proved to be quite brittle. After placing them in a sample bag, he dug down and uncovered more of the fibers. The rods penetrated deep down into the ground.

"Fibers form branching network around crystals," Boris said. "The deeper I dig, the denser they become. I do not know how far down they go, but I think maybe quite a way, maybe many meters."

He then took out the mechanical probe from his pack, pounded the pointed end into the ground, and turned on the control that activated the powerful motor behind the tip. The inner part of the probe began to spin in its casing, digging down and disappearing into the ground. Holding the outer casing in one gloved hand, he fed the connecting cable with the other hand as it followed the tip in its downward course.

"Temperature gets higher as probe goes down. That means energy being produced, maybe related to EM radiation we've picked up."

He reversed the direction of the spin and pulled the cable up as the inner part of the probe screwed itself back up to the surface.

"Next time we bring portable X-ray to get deeper look underground," Boris said.

After chipping and bagging more samples, he hopped over to another area and found the same thing: crystalline rods connected to each other by fibers, with the whole mass penetrating deeply into the ground.

"This reminds me of a mushroom network on Earth," Samantha said, "with the fibrous mycelia extending throughout the soil, connecting one mushroom to another."

"Boris, what do you think the crystals and fibers are made of?" Kilborn asked.

"I do not know, Captain, but maybe like rocks on surface they are silicon-based. I need to examine samples back on *Hermes*."

While he filled his bags with samples, Akira took some pictures. After nearly 2 h, the six of them returned to the rover, where they radioed their discovery back to Tilda and Anthony. Famished, they unpacked their dinner and began to eat.

"We should rest for 30 min or so, then start our return back to the ship," Kilborn said.

"Go, already?" Akira asked. "But...this is such an unprecedented find! What is this doing here? This network has spread over part of this plateau and is somehow generating ELF radiation. It looks like an active process. We need to understand what's going on."

Kilborn looked sternly at him.

"I know you'd like to stay longer to study this thing, but I'm concerned about the EM levels. Our suits seem to be handling things OK, but we shouldn't press our luck. Plus, we have our samples and photos to analyze, as well as our instrument recordings. We can study this thing better and safer back at the base."

"But Captain..."

"That's enough, Akira. We'll rest a bit and then return to the *Hermes*. We'll have time to evaluate our situation and consider what to do next. We'll likely come back soon and explore further."

Glaring at the Captain, Akira remained silent.

Akira sure doesn't like this plan? Samantha thought to herself. *But I can understand his feelings. This is a major discovery. We need to understand what's happening. We've never seen anything like this before. But why doesn't the Captain want to stay longer? The radiation is stable, and we're doing fine. What's his concern?*

She kept her thoughts to herself, and after their rest they headed back to the *Hermes*.
****

It roused from its slumber for the billionth time. It became aware again. It remembered again as it basked in the glorious light from above.

"Hail to Maker, life-giver!" it rejoiced. "Must find sustenance deep down. Must grow."

It became aware that a small part of it was missing. Pieces had been broken off. This was puzzling but not unprecedented, since sometimes objects from above came crashing down to the ground and occasionally caused it some damage. Once before, it had destroyed a large object flying overhead before it came down, but that was because the object was radiating energy it could understand.

"Must become whole again," it considered.

But this would be difficult, since the broken parts were close to the surface, where the temperature was too cool to allow for the crystal-forming process to occur. Perhaps the missing pieces could be compensated for by crystals forming and pushing up from below. It was there that additions could be made in the concentrated heat during the hot time, when the Maker was above, and the cooling off of the cold time, when the Maker left and it went to sleep.

"Yes, growth and repair will occur," it thought.

Mercury, Day 42

The next day was spent analyzing the pictures and data samples collected from the network. That night, after dinner, the crew gathered to discuss their findings. Boris began.

"Analysis shows giant crystal rods exposed to Sun are pure monocrystalline silicon, like what is used for solar panels. Fibers connecting them are highly pure silica glass, like in fiber optics."

"Any idea how such a system like this can exist in this hot, dry environment?" Kilborn asked. "Don't you need water to form crystals?"

"Well, in semiconductor industry on Earth, silicon rods made by pulling seed crystals through bath of molten silicon, then cooling them. Maybe something like this happening here."

"What do you mean?"

"Probe shows temperature rose the deeper it went, and pressure too. Maybe hot molten material is below plateau surface. Sunlight refracting down through crystals could be adding to heat. At some depth temperature and pressure high enough for molten silicon to form subsurface rocks. Maybe when Sun sets during local nighttime, things cool off. New crystalline material is added to bottom of rods. High pressure also could push rods up, breaking surface."

"My God, this is astonishing, if true."

"Yes, Captain. Is best idea I have. We need to go back and do more tests. Portable X-ray will help us look below."

Turning to Akira, Kilborn asked: "How might this relate to the signal?"

"With the crystals absorbing the light pouring down from the Sun, they may be producing heat energy and possibly a current that is being transmitted throughout the fiber network. Also, if there is molten silicon down below, this could play a role. Where you have ion-containing molten material in motion in a gravitational field like on Mercury, you can get EM waves."

"But why here on that hilltop? What's special about it? How can there be molten material below the ground."

"I don't know. But I agree with Boris—we need to take a closer look at what's going on below the crystals."

"I have a question," Tilda asked. "I didn't pick up any signal until yesterday. If there was already hot molten material underground, why did the signal start to occur then?"

"Let me check something," Akira responded.

He consulted the computer watch implanted in his wrist, punched a few buttons, then continued.

"Depending on the pressure and composition of the material, silicon compounds melt at around 1400 °C. So if Boris is right about some kind of heat source underground, if we assume that it can produce a temperature of 1100° or so and that the crystals can concentrate the Sun's light to match or exceed an outside temperature of 300°, then this would reach the critical temperature needed to melt some of the silicon-containing rocks located under the network. In addition, the pressure generated in this environment is likely to be very high, assisting the melting process."

"But what can produce a temperature of 1100°?" Samantha asked

"Molten lava can reach that temperature," Boris answered. "Maybe lava is under network, and hill we climbed is quiet volcano. Maybe lava tubes exist to move lava up to network."

Chuck's eyes widened.

"Whoa!" he said, "and we were right on top of it. It's a good thing that it didn't pop off."

"So how did the crystals get there to begin with?" Kilborn asked Akira.

He thought for a moment, then said with a straight face: "Perhaps alien explorers visiting Mercury left behind some silicon chips from their computers before they went home."

Everyone laughed.

"I know this sounds far-fetched, but the existence of the network itself is far-fetched. An external source placing the crystals on Mercury needs to be considered."

"Any other possible explanations?" Kilborn asked.

Akira thought for a moment.

"I guess a more home-grown explanation could go something like this. Eons ago, a small asteroid containing silicon matter crashed on top of the network's volcanic hill during local daytime. This crash released molten lava that bubbled up to the surface, and together with the heat from the Sun, some of the adjacent rocks were melted into pure silicon. During the subsequent night-time, this hot silicon mass cooled and crystallized. Once the process started, it continued and grew with each Mercurial day-night cycle. Over time, the network came into being."

"There's a lot of conjecture here," Kilborn said, "and a large mass of crystals to consider."

"We have several billion years to play with, Captain. That's plenty of time for a mass the size of the network to develop, little by little, with each solar passage."

"We need to go back," Boris said. "We have more tests to do."

"Yes, Captain," Akira continued. "There are likely more secrets to uncover."

"All right," Kilborn said. "You two can go back tomorrow to run your tests. Chuck will drive, and Samantha will go with you. But be careful. We'll monitor things here and warn you if there are any signs of an increase in the intensity of the signal. We wouldn't want that to happen when you are up on the plateau."

Mercury, Day 43

The exploration party left early the next morning in *Rover 2*, taking along the portable X-ray machine and some related equipment that had been stored in the ship's cargo hold. They arrived by noon, parked the rover at the foot of the hill, and ate a brief lunch. While Chuck again remained in the vehicle, Akira, Boris, and Samantha lumbered up the slope, pushing and pulling the X-ray machine on its rollers. According to the comm from Tilda, ELF and EM waves of other frequencies continued to be broadcast, but at an intensity insufficient to penetrate their space suits. At the top of the hill, Akira and Boris set up the X-ray imager near the periphery of the network. While Akira began to scan the sub-surface, Boris drilled a heat-sensitive probe into the ground nearby.

"I show the rods and fibers penetrating down into what looks like a large cavernous space," Akira said. "The density of the space is consistent with a thick fluid, and I see ripples suggestive of movement."

Boris then spoke after consulting the readings from his probe.

"Da, I am also in cave area. There is resistance to probe, so supports presence of fluid and high pressure. Temperature reading is over 1100°, so consistent with lava-like material."

"Right on!" Akira said. He continued to gaze at the X-ray. "As I make adjustments to observe even deeper, several vents appear to lead down from the cave to below the surface level of Caloris, possibly to a magma-filled reservoir even deeper down. Perhaps the source of the subsurface heat may come from an area close to the planet's silicate mantle."

"Fascinating," Boris said, retrieving the heat probe from the ground. He turned to Samantha, pulling out and starting his portable saw.

"I want to take more samples of network. Samantha, go back please to rover to get heavy duty saw blade so I can cut into crystal rods."

As she turned, she noticed him beginning to cut some of the network fibers. She walked over to the edge of the plateau and began her descent. When she was about a third of the way to the rover, the ground shook, nearly causing her to lose her balance. She felt a tingling in her body and a curious emotion that was similar to fear. Then the shaking and the unusual feelings stopped.

What's happening? She thought.

She heard a voice in the suit's intercom.

"This is Chuck. What's going on up there? Are you all right? Tilda radioed to say that she just recorded a strong EM pulse from your area that included a lot of ELF waves."

"Chuck, this is Samantha. I'm on the path leading down from the plateau. Akira and Boris are above. I'll go up and check on them."

She turned and bounded up the slope, radioing for Akira and Boris, but receiving no response. As she climbed onto the plateau and ran toward the network, she spotted the prostate bodies of her companions. Shocked, she ran over to Akira and then to Chuck, the saw still in his hands. She scanned both of them with her vital sign monitor, and each was dead. Peering through the face plates of their suits, both looked burned, and their intra-suit temperatures were very high.

"Chuck, both Akira and Boris are dead," she announced, shaking off her anxiety. "I think they've been thermally fried, possibly by the ELF radiation that Tilda picked up."

After a brief silence on the other end, Chuck said: "Samantha, I'm relaying the comm link from the *Hermes* to you. Go ahead, Captain."

Kilborn's voice came through Samantha's audio link.

"What the Hell happened out there?" he bellowed.

"This is Samantha, Captain. I don't know for sure. I was heading back to the rover to get a heavy duty saw blade. Boris wanted to cut a sample of a crystal rod to bring back to the ship."

She glanced over at the network and saw some pieces of cut fiber near a gouge in the side of one of the crystalline rods.

"He must have cut into one of the rods. Maybe the network reacted to this by emitting a strong EM pulse. Captain, both Akira and Boris are dead, and they show signs of extreme thermal radiation."

After a silence, Kilborn ordered: "You and Chuck grab the bodies and get out of there. You must abort your mission. You can give us a complete report when you return."

"Aye, aye, Captain," she said.

Ten minutes later, Chuck appeared over the edge of the plateau—he had left the rover to give assistance. Together they pulled the bodies of Akira and Boris down the slope and into *Rover 2*. Leaving the heavy X-ray machine behind, they solemnly began their journey back to the *Hermes*.

****

It became aware that more pieces of itself were being removed. But there was no indication that this was due to an object crashing down from above. Rather, there was something moving around its fibers that was damaging them. Then, something began to attack one of its rods. This had never happened before. It became alarmed. It experienced fear.

"What is happening? Must stop damage."

A general arousal reaction occurred in the only way open to it. All of its thought waves concentrated into a general EM pulse that was discharged in the direction of the intruder. The intensity of this focused discharge was increased several fold over what it normally would have been at this time of the Maker's transit overhead. When it was over, the destruction had stopped. The source of the damage had been terminated, and it could go back to growing itself with the help of the Maker.

****

Samantha spent most of the trip back in deep thought, reflecting on her experiences and thinking how she would report them to the Captain. Her thoughts had only been interrupted when she took time to put on her filtered glasses to look at the Sun, which due to the odd orbital characteristics of Mercury had appeared stationary in the sky. When they had reached the *Hermes,* Chuck drove the rover to the base of the ship, and they unloaded the two bodies and put them into the outside elevator. Then they got in, and all of them ascended to the airlock. Leaving the bodies in the elevator, Chuck and Samantha entered the ship, removed and stowed their space suits, and took the inside elevator up to the mid-deck, where the rest of the crew was waiting for them.

"Welcome back," Kilborn said. "Grab some coffee and let's talk."

They gathered around the food table. Kilborn bent over and activated the recorder transmitter lying on the table that was connected remotely to the central computer.

"This is Captain William Kilborn. It is 18:35 Earth Standard Time on Day 43 of the Mercury expedition. This is a continuation of my comments from earlier in the day. Engineer Morgan and Astrobiologist Evans have just returned from their excursion and are ready to give their report on the death of the two crewmembers and the events concerning the network that led up to it."

He motioned for Samantha to speak into the recorder microphone. She reiterated what had happened on the plateau. When she had finished, Kilborn moved closer to the mic and looked at her sternly, as if he were a prosecuting attorney in a courtroom.

"Dr. Evans, what was the cause of death of your two crewmates, in your opinion?"

"I think the powerful EM surge from the network penetrated their space suits. I understand that the *Hermes* recorded the pulse as primarily consisting of ELF waves. This would support my observation of the bodies that the cause of death was an intense thermal event."

"And what do you think prompted the radiation surge?"

"I think it was what my crewmates were doing to the network. Everything seemed fine until Dr. Baganov began to saw away at one of the rods. I think this triggered the response that resulted in the deaths."

Kilborn paused, then continued his interrogation.

"And you were not affected because you were part way down the slope leading back to *Rover 2*, correct?"

"Yes, Sir, although I felt the ground shake and a tingling sensation. I believe the tingling was a non-thermal reaction that was transmitted through the rock between the plateau and my position on the side of the hill. Luckily, the rocky mass attenuated the ELF waves to the point where I was not killed myself."

"Yes, that is fortunate, Dr. Evans."

"But there's something else. I distinctly experienced a fear emotion that was not my own, one that was transmitted to me."

Now I've done it, Samantha thought to herself. *Time to open up the can of worms.*

All of her crewmates were staring at her, except for Anthony who was looking down at the floor. Lars appeared concerned and leaned in toward her. She continued.

"I think the emotion came from the network. I think that it experienced fear and transmitted this experience to me. I think the network is a conscious life form."

"What!" Kilborn said, standing stiffly upright. "What are you saying?"

"I've been thinking about this issue all the way back from the plateau. I just don't have any other explanation."

"How do you know that the fear emotion was not simply you reacting to the event?" Kilborn continued.

"Because the fear response didn't feel like me. It seemed foreign, like something else was putting it into my head. I actually was afraid at the same time, but this emotion was different—it was me, not someone else."

"Well, that certainly is a lot to take in! I think we had best conclude the report for now until you've had a chance to eat and rest and be examined by Dr. Ahlgren. We can continue this at another time."

"I'm not crazy, Captain. This is what I experienced and what I believe to have happened," Samantha persisted.

"I'm not accusing you of anything. But I want to talk further with Mr. Morgan and with you after you've had a chance to put some distance between yourself and the tragic episode. So, I declare this recording to be terminated for the time being."

He turned off the transmitter.

Mercury, Day 44

The ghost-like figure came toward her. Samantha backed up against the rock wall. She pulled out a laser pistol and fired, but the intense beam simply went through the approaching creature, failing to stop its advance. It was upon her. She felt blistering heat and smelled smoke. The apparition enveloped her as it took on an ape-like form. Its mouth opened to reveal shimmering fangs. They penetrated into her skull. The creature began to eat her brain. . .

Samantha woke up with a start, her heart pounding like a drumroll in her chest. Her sheets were soaked with sweat. Gazing out the view port next to her bed at the myriad stars started to relax her. She reached out her hand, triggering the wall sensor, and the light went on. She scanned the small room and surveyed her domain: the storage bins above and below her bed that contained her clothes and personal items; the swivel chair and pedestal desk, on which sat her personal computer; the 3D pictures of her parents and Solar System bodies that decorated her walls. The familiarity of these items seemed to ground her and take away the residual anxiety that had lingered after her nightmare.

Rubbing her eyes, she sat up, slipped on a night shirt that extended down to her mid-thighs, and glanced at the desk clock. It read 01:33. She went to the door next to her desk, opened it, and went out into the hall. Circling the central elevator were the other seven sleep pods and the two bathrooms. Turning to the left, she went into Bathroom A and sponged the sweat off of herself. When she returned, she heard the sound of a computer keyboard coming from Lars' pod, which was adjacent to hers. On impulse, she gently knocked on the door.

"Yes. . .come in," his voice said from within.

Opening the door, Samantha saw that he was seated before his computer, still dressed in his daytime solid blue jumpsuit, the standard apparel for crewmembers.

"Sam, how are you?" he said, motioning for her to sit on his still-made bed.

"Not great, Lars. I had a horrible nightmare, something about a monster eating my brain. I just can't get over what happened yesterday."

"Yes, it was tragic. After Chuck and I retrieved the two bodies and put them in the cold room for transport back home, I performed a mini-autopsy. I think you were right in saying that the two of them died from thermal penetration due to the ELF waves generated by the network."

"What a way to go," she said, blankly staring at the floor. Then, recovering, she looked up at him.

"Lars, I don't know what happened to my mind. Something penetrated it. I felt a foreign emotion. It was one of fear, but it wasn't me. Am I going crazy?"

He thoughtfully considered what she said.

"I don't think so. But I don't know how to explain your experience. In fact, I was just searching through the ship's computerized data base looking for a possible explanation when you knocked."

Her eyes widened.

"Did you find anything?"

"Nothing specific in the medical literature. Nevertheless, I became intrigued by the similarity between our human brain waves and the ELF waves transmitted by the network. Both operate at a similar low frequency, less than 45 Hz. The difference is that the network's waves are much more intense than brain waves. But since wave forms of similar frequencies can resonate with each other, I'm wondering if something like this happened to you. Maybe the network felt fear associated with the cutting of one of its rods, and this was communicated to you by transmitting waves directly into your brain's limbic system."

"But why wasn't I killed?"

"The strong pulse that killed Akira and Boris likely was directed to both of them. It was intense and largely thermal. You were below the level of the plateau and shielded somewhat by the intervening rock. Plus, the winding path down put you on the other side of the hill from where your crewmates were working. Whatever reached you produced more subtle non-thermal effects. I admit, this explanation is a bit speculative, but I don't know how else to account for your experience."

"So maybe the network really is a conscious entity."

"There's a school of thought that says that consciousness is produced by electromagnetic radiation. . .I read about this theory earlier tonight. In humans, the aggregate of our neurons firing in synchrony results in brain waves that are experienced by us as consciousness. But you don't really need neurons, according to this theory. Anything that produces a synchronous wave front can result in a conscious experience. So I suppose that this would even encompass molten silicon ions in motion, such as what might be occurring below our friend on the plateau. Perhaps these waves are being transmitted throughout the network."

"Then I guess it would stop being conscious when things cool off a bit."

"Possibly. It certainly would explain why we didn't receive a signal until the temperature increased to some critical value: it just wasn't hot enough to wake the network up! It also may account for why the signal waves are largely in the ELF frequency range—this is the predominant frequency of the network's thought processes."

Samantha paused to consider what he was saying.

"So. . .we still have another 100° to go before reaching Mercury's maximum daylight temperature. Maybe the network will become even more aware and smarter over the next couple of weeks."

Lars laughed.

"Maybe. But of greater concern to us. . .the ELF waves could become more intense. We know that they can be detected from space. Who knows what they'll do to our ship—we aren't that far away."

She shivered.

"My god, that's a scary thought."

"Well, these ideas are quite speculative."

"But could we be in danger?"

"I don't know. We have a strong heat and radiation shielding system on *Hermes*—enough to deal with the rays of the Sun. That might be strong enough."

He reached out to take her hand.

"At any rate, let's not worry about this happening until we get some evidence for it. But I plan to discuss some of these ideas when we have our next meal discussion period at lunch later today."

He smiled again and looked into her eyes. She relaxed and felt comforted. But she also felt aroused. She shifted toward him, the hem of her shirt riding higher up on her thighs. He glanced at her legs, then got up and moved next to her on the bed. He put one arm around her waist and caressed her hair with the other. He kissed her lightly once, then more passionately a second time. She responded with enthusiasm as the two of them reclined on the bed. Their friendship progressed to a more intimate level.

****

Later that morning, the crewmembers gathered for a brief memorial service for their two departed crewmates. The service was holographically recorded and broadcast back to Luna City. The cover story was that Akira and Boris had died in a cave-in while exploring some geological features under the ground. The Space Council had ordered them to resist mentioning anything involving the network until more information could be gathered concerning what had happened. Samantha did not like this order, which she thought was unnecessarily deceitful.

At lunch later in the day, Lars explained some of his ideas about the event that killed their two crewmates, and he echoed Samantha's notions and the validity of her experiences. Kilborn asked her what she thought.

"I discussed these issues earlier with Lars," she said, omitting the details of their interaction afterward, "and I think they should be taken seriously. I don't know how else to explain what happened and how I felt during the event."

"I see," Kilborn said noncommitingly.

"But I have a question, Captain." she continued. "Why aren't we telling people throughout the Solar System about the network and what really happened?"

Kilborn looked at her.

"That is a difficult question to answer. . ."

"Give it a try, Captain," said Lars.

Kilborn sighed.

"OK, I'll level with you. Because these aren't the first human deaths caused by this. . .thing."

"What!" Samantha said.

Chuck, Tilda, and Lars also appeared to be stunned. Anthony looked away, as if embarrassed.

"Shortly after a third flyby probe recorded the signal, the Space Services Council decided to send a manned Space Navy scout ship to Mercury to investigate. It was all top secret. No one knew if this was a real signal produced by a device placed on Mercury by some alien group, or if it was a weapon put there by the Mars Secession terrorists, or what it was. The scout ship reached the planet when Caloris was at its highest daytime temperature. It went low toward the surface, taking pictures and measurements. Nothing new was detected, except that suddenly there was a surge of EM radiation largely in the ELF range that penetrated the scout ship's electronic radiation shields and destroyed the computer software. The ship managed to fly on for a while, but ultimately it began to break apart and crashed about 120 km southwest of us."

"Why weren't we briefed about this?" asked Chuck.

"As Space Navy officers, I knew, and Anthony knew, but it was decided to keep the rest of you in the dark—even Tilda, who is in the Army. The reason? Partly to keep things secret. Some of the military brass still believes that there is sabotage or a security issue involved. But partly to see what conclusions our science team would come up with on its own. Our mission was always seen as a scientific one—that's why the mission included you scientists and engineers. Anthony and I were ordered to ferry you folks to Caloris, try to find out what was going on, and then get out of here. No one knew about the crystals."

"But didn't we have a right to know about the danger?" Lars asked.

"I suppose so. But both Anthony and I are under military command, and so we had to follow orders."

Tilda glared at Kilborn.

"I guess knowledge of this information is beyond my pay grade," she said sarcastically. "So, what now...Sir!"

"We wait and monitor what happens: to see if the ELF waves increase in intensity as the temperature goes up; to see what this crystal network does next; to pay it another visit or two. If the situation begins to endanger this mission, then my orders are to abort it and skedaddle home."

The group was silent.

"OK, then," Kilborn continued, "let's finish our meal and get back to our maintenance schedule. It's been disrupted by events. Tomorrow, we'll consider our next course of action as regards the network."

Mercury, Day 45

The next day at breakfast, they discussed their options.

"Here's the situation," Kilborn said. "The network is radiating ELF waves at an acceptable intensity, so long as we leave it alone. Akira and Boris got in trouble when Boris tried to cut into a rod, so we don't want to do that again. The question is, was the response an intentionally hostile action?"

"I don't think so, Captain," Samantha said. "I felt fear from it. I think the surge was kind of a knee-jerk reaction, a kind of hyperalert, stimulus-response reflex that resulted in increased ELF radiation that was intense enough to kill our crewmates. It was trying to protect itself from further damage."

"I agree," Lars added. "There's no indication of a conscious plan to harm us."

"OK, so if we don't provoke it, it won't react in the same way," the Captain said. "Our orders are to study this thing, so long as the mission is not in danger. We can do this from the ship."

"Sir, I'd like to go back and examine it," Lars said. "If the network is conscious, it would be the only non-carbon thinking entity that exists in our Solar System. The fact that it may be adding more crystal matter to itself—feeding itself, if you will, and growing as a result—adds to the notion that it's some kind of living entity."

"I'd like to go with him," Samantha said. "We've never found conscious alien life before, and this is too exciting to pass up."

Kilborn thought a moment, then turned to Tilda.

"How is the radiation doing?"

"The intensity is increasing, but slowly, Captain. . .as the temperature rises. But it's within tolerable limits right now."

He turned back to Samantha.

"So you think you two can go back to this thing and survive?"

"Yes, Sir," she said. "It was the surge that killed Akira and Boris. The baseline radiation is trivial by comparison. We can only detect it because of the sensitivity of the ship and satellite receptors."

"Lars, what do you have in mind if you get up close and personal with this. . .entity?" Kilborn asked.

"We have an electroencephalographic machine on board to monitor human brain waves. Chuck tells me that he can adapt the electrodes and receiver of the EEG to measure the variations in the ELF waves of the network. I want to see what this will show us. Maybe we can confirm that it truly is conscious."

"How do you plan to do this?"

"We would expose it to some varying stimulation in terms of electromagnetic frequencies. We have a portable EM frequency generator on board the ship that could do this. We could set it up on the plateau, wire the network to the EEG machine, and see how it responds."

"Wait a minute," Anthony said. "We don't want to trigger another surge."

"No, we wouldn't cut into it or crank up the EM intensity very high. We certainly don't plan on dying. We would just produce weak but variable degrees of external stimulation and measure the resulting responses."

Kilborn looked thoughtful. "I don't know about this plan. . ."

"Captain," Samantha interrupted, "isn't our mission to find out what caused the signal? We might be able to fulfil this mission with the plan being proposed by Lars. Plus, think about the fact that we might confirm the existence of a conscious entity in our own Solar System, one that's silicon-based."

Kilborn thought for a moment.

"OK, let's see how high the radiation is tomorrow morning. If it's not too bad, you two and Chuck can take *Rover 2* and pay another visit to this thing on the plateau. But if there's any indication that it's becoming a danger, you get out of there. And if there's any threat to our ship, I'll order the mission aborted and take off for home, whether or not you're back."

"Understood, Sir," Lars said. Samantha nodded in agreement.

After a moment, Chuck said: "Count me in as well!"

Mercury, Day 46

The Captain approved the excursion the next morning after Tilda reported stable radiation levels. Chuck, Lars, and Samantha suited up and took off in *Rover 2* right after breakfast. They took along the EEG machine, which Chuck had modified the night before, and the EM generator.

The trip to the network was uneventful. Overhead, the Sun had resumed its westerly course as it continued to beam down its blistering rays onto Caloris.

"There's the peak, up ahead," Chuck said. "We'll be there in 10 min."

"Let's review our plan," Lars said. "First, what exactly did you do to the EEG machine, Chuck?"

The engineer responded as he steered to avoid a large rock in front of them.

"Since the network generates its ELF waves at a much greater intensity than human brainwaves, I had to crank up the resistance and modify the leads to allow us to track the waves without overwhelming the electronics."

He pointed to a container of heat-resistant adhesive at the back of the cabin.

"Use this to place the leads directly on several crystal rods spaced far apart, and then just start recording. I have a diagram of a possible placement pattern etched onto the body of the EEG machine. Like all of our other tools and equipment, the metal alloys in the machine and wires can resist temperatures of up to 500 °C, so the heat won't be a problem. You should have no difficulty monitoring the waves that are produced by this. . .crystal brain."

Samantha cringed at this description. She turned to Chuck.

"We can't say for sure that it's a brain. All we know is that the network produces ELF waves and may be conscious. We don't know if it can think logically, or appreciate a ballet, or want sex."

"OK, Sam," Lars said. "We set up the EEG machine, I'll monitor it and record what we get as a baseline, and you use the generator to broadcast a variety of EM waves in terms of frequency and intensity in order to see how the network responds."

"Got it," she said.

When they reached their destination, it was around noon time. After eating lunch, Lars and Samantha carried the EEG machine up to the plateau, along with the portable EM generator. Passing the earlier discarded X-ray machine, they approached the network. They hooked up the EEG leads according to Chuck's diagram, and after making some adjustments, ELF waves appeared on the monitor.

"There they are," Lars said. "I'm showing waves of about 10 Hz. If it were a human, he would be pretty relaxed."

"And I'm starting to feel a tingling in my skin," Sam said.

"Even through your suit? Wow, you must be very sensitive to these waves. In studies of people in laboratories, only about 10 % can sense ELF waves. They didn't have space suits, but the intensity was much less, about 5 kV/m, so it's hard to compare with our situation. But still, I don't feel anything."

"Let's try the generator."

She activated the EM generator and set it to broadcast at 50 Hz at the minimum baseline level of power.

"There's no change in the network," Lars said. "Cool and calm at 10 Hz."

"All right, let's try radio frequencies."

She increased the frequency a thousand-fold, to 50 kHz.

"The network's getting a bit activated. It's up to 14 Hz now."

At 50 mHz, the EEG showed more activation, to 22 Hz.

"OK," Lars said, "let's try generating more power."

Samantha began to increase the amplitude of the generated waves.

"Whoa!" Lars said. "That got a reaction. The network is now showing 30 Hz waves. That would put it in the beta range of the human EEG—very alert and activated. And the amplitude of the waves is rising as well, to levels that are beyond the measurement abilities of this machine."

"And I'm starting to feel worried, anxious. It may be what the network is feeling in reaction to the increased intensity of the stimulation."

She then cut back the power but continued to increase the frequency. The network became less activated when they hit the mid-microwave range.

"So," Lars said, "to sum up, the network is more sensitive to radio frequencies than to higher microwave frequencies, and it reacts more to higher versus lower intensity waves."

Lars checked the computer watch embedded in the top of his left hand glove.

"We've been out here long enough. Let's call it a day."

Samantha shut down the generator.

"That did the trick," Lars said. "The EEG is showing 10 Hz again."

"And I'm feeling sad," she responded, "kind of lonely."

"So assuming that you're in empathic contact with the network, it would appear that it's able to react emotionally to outside stimulation. It can go from quiet attention, to hypervigilance and anxiety, to some kind of despondency."

They secured their equipment and made their way down the slope to the rover. As they began to head home, they briefed Chuck on what happened.

"That may explain the antenna we found," Chuck said.

"How so?" Lars asked.

"The flyby probes and the Space Navy ships all broadcast in radio frequencies. The network may have picked up radio signals from passing ships, become activated or anxious or whatever, and reacted by sending out a pulse of high intensity ELF thought waves that could be monitored from above."

"And the antenna?" Lars asked.

"Well," Chuck continued, "in the case of the military scout ship that went closer to the surface to investigate, the ELF waves could have penetrated its electronic shields, maybe even damaged its computers, and caused it to break up. The antenna and adjacent solar cell fragment was a product of this destruction."

"That makes sense," Lars said.

"I don't know for sure that this happened," Chuck continued. "I'm not an expert on Space Navy ship construction. I'm just an engineer contracted for this mission, and I've only flown a few military flights. I'm not knowledgeable of the class of the ship that was sent here—the Captain didn't say, it being a secret and all. But still, space vehicles tend to be built the same. What I suggest is certainly a possible scenario."

"I wonder what other secrets the Captain is keeping from us?" Lars asked.

Samantha turned around to look at him.

"What do you mean?"

"I mean that he withheld some important information earlier that affected our lives. This signal and its cause seems very important to the military. With the Secession movement on Mars, the Earth and Mars governments are very nervous, and so is the Space Council. This naturally makes the Space Services military people jumpy. Who knows what other orders Kilborn has received that he's not telling us about."

"Yeah," said Chuck. "I'm wondering the same thing. This is a science mission, according to the Captain, and the *Hermes* is not supposed to be capable of undergoing a military action. The press was all over the mission before we left, so it was a highly visible operation. Still, the Captain, Anthony, and Tilda are all military personnel, and you wonder what might be up."

"Anyway," Lars said, "there's not much we can do about our suspicions now. We need to go back and brief the rest of the crew about what we discovered today. We seem to have a reactive, maybe conscious, alien presence to deal with that can be deadly but doesn't appear to have malicious intent."

"Well, there is some positive news," Samantha said.

"What?" asked Lars.

"We haven't seen any Mars Secession terrorists hanging out around Caloris."

The two men laughed, but much of the rest of their return home took place in thoughtful silence.

****

It became aroused. Something was occurring nearby. Energies were being sent to it, some strong, some weak, some barely sensed. This varying stimulation was something new, something unexpected, for it had never happened before in the billions of years of its existence. It found itself reacting in a variety of ways.

"Something here. Not me. Not Maker. Not sent down from above. Not destroy me—just affect me."

The newness frightened it. It became more alert. It found itself beginning to react, when suddenly the stimulation stopped.

"No more stimulus. Alone again with Maker."

It relaxed. But something else was happening that was new. It became curious about what was happening. It wanted to know more. Something other than the Maker existed.

"Why did it stop? Why did it leave?"

Also for the first time in its existence, it felt alone.

Mercury, Day 48

To: samantha.evans@Mercuryexplore.assc (RESTRICTED USE)
From: k.evans@stationdelta.org
Subject: Hello
Date: Mercuryexplore, Day 48
    Dearest Samantha,
    I have very bad news to tell you. Your father suffered a stroke and has been in a coma ever since. In the past few weeks, he lost a lot of weight since he hadn't been eating much, and he became very weak. This caused him to prefer to sit alone and stare blankly at the holograms being projected in front of the wall across from his bed. The staff tried to get him to respond to the music therapy, but whenever they tried, he got angry and lashed out at them. As you know, this is not like him, to be cranky and irritable. It must have been his Alzheimer's. I've been told that this can happen to patients with this disease when the frontal lobes of their brain lose their moderating effect on emotions, or something like that. Anyway, his heart had been beating irregularly, and the doctor's think he threw a massive blood clot that traveled into his brain. We are carefully watching him, but no one is optimistic. I will keep you informed.
    Love,
    Mom
    ****
To: k.evans@stationdelta.org (RESTRICTED USE)
From: samantha.evans@Mercuryexplore.assc
Subject: Dad's stroke
Date: Mercuryexplore, Day 48

Mom,

I'm so sorry to hear about Dad. He was doing so well, at least in terms of his mood and activity level. Let's hope that he comes out of his coma. How are you doing? It must be a terrible stress for you. I'm sorry that I can't be with you at this time. Just know that my thoughts are with you and Dad. I hope that John Gomez and your other friends have been supportive.

We've found the cause of the mysterious signal. It appears to be a clump of silicon-based crystals and fibers that have developed on Mercury and are able to use the heat from the Sun and an underground magma source to generate low frequency electromagnetic waves. We're trying to figure out how it formed and how it is able to make these waves. I will let you know more when we have some answers. Keep this information to yourself. For some reason, the mission operations people don't want it getting out to the public at this time.

Love,

Sam

Mercury, Day 52

To: samantha.evans@Mercuryexplore.assc (RESTRICTED USE)
From: k.evans@stationdelta.org
Subject: Sad News
Date: Mercuryexplore, Day 52

Samantha,

I'm so sorry to tell you that you father passed away yesterday. His stroke was too massive, and he was so weak from his weight loss that his heart just gave out. I'm making plans for the funeral, although most of the arrangements were already in place after he went into his coma. It's unfortunate that you can't be here, but I know you will be here in spirit. You know that he always adored you and was proud of your accomplishments. Me too. You have been a source of comfort and pleasure to us throughout your life.

I'm doing fine with the news. In a way, your father has been dying to me for the past several years, and I've learned to cope with the inevitable. My friends have been very supportive. John Gomez especially has been kind to me. We've had dinner together nearly every night the past few weeks, and he has been there for me when I've felt blue. Right now, he's helping me with the funeral arrangements.

Your discovery of the crystal entity is very exciting. Do you think it's some kind of life form? We seem to be finding little creatures all over the Solar System. This sounds like something new, being silicon-based. You must be having a field day trying to figure it out. I'm sure that you and your crewmates will unravel the mysteries surrounding it in due course.

Take care. And don't worry about me here. Things are moving along, and I have plenty of support.

Love,

Mom

****

Samantha read this message after lunch, tears forming in her eyes. She went to the aft-deck and knocked on Lars' door, where she knew she would find him reviewing laboratory data on his computer.

"Come in," he said cheerfully, looking up from the terminal. He became serious when he saw the tears. "Sam, what's up?"

"It's my father. I just heard from my mother that he passed away. His heart just gave out due to the stroke and his weakened condition."

Lars stood up, walked over to her, and gave her a hug.

"I'm so sorry for you and your mother," he said. "I guess it's no surprise, and he likely passed peacefully and without pain. But I imagine that you feel terrible nonetheless."

The tears continued to stream down her face.

"I do, Lars. I always had a special relationship with him."

She sobbed a bit, then continued after regaining her composure.

"He encouraged me to go into science, even though that wasn't his field. He was always there when I had some sort of problem. I've felt guilty about being away so much, from both my father and mother, and now here I am, millions of miles away, and I can't be at his funeral. I feel horrible, and my mother is alone, although she seems to have friends supporting her, but I still feel. . ."

"Shhh," Lars said. "Sit down here on the bed." He pulled a tissue out of the container on his desk and gave it to her. She wiped her eyes.

"Thanks, Lars, I guess I got a little carried away"

"That's OK. You should have seen me when my wife died. It's hard to lose a loved one when you're away. I was on Luna City when Laura died on Earth, and I felt badly that I couldn't have been with her in the air car. Maybe I could have intervened to prevent the accident. Maybe we would have done something that day to avoid even being in the vicinity of that drunken driver. It's easy to feel guilty when someone you love dies."

"I know you're right, and I guess I'll feel better with time. I really appreciate your being my friend right now."

She gave him a hug, and the two of them embraced for what seemed like several minutes.

"Well," she said, "I guess I had better get back to the lab. I'll contact my mother again tomorrow morning after I have a chance to settle down a bit."

She paused to blow her nose, then quietly left.

Mercury, Day 53

To: k.evans@stationdelta.org (RESTRICTED USE)
From: samantha.evans@Mercuryexplore.assc
Subject: Condolences
Date: Mercuryexplore, Day 53

Mom,

I was very saddened to hear about Dad's passing. I guess it was no surprise, but that doesn't make the pain any easier to deal with. Like you, I have a special friend here, Lars, who I can talk to and whose shoulder I can cry on. Since he lost his wife in that terrible air car accident, he understands the grief of losing a loved one. I'm glad that John and other friends of yours have been supportive. Let me know if there's anything I can do about the funeral—maybe send a eulogy message that someone could read, or something. Obviously, I'm limited being so far away. But keep me in mind—I want to be a part of Dad's service.

The crystal network we found has indeed been interesting. No one knows yet what we're dealing with, or if it is some sort of living entity. Time will tell. I'll let you know what we find.

The temperature here continues to rise, but so far the electronic radiation and heat shields in the ship's hull seem to be dealing with things, both from the network and from the Sun. It's amazing how powerful the Sun is! Being so close to it really makes you aware of its power.

Actually, some of us think that the network is able to harness solar energy for its own benefit. If so, this would be another example of Darwinian adaptation in action. Like the organisms on Mars, Europa, and Titan—and on Earth, for that matter—the network seems to have evolved ways to not only cope with its environment, but also to benefit from it. We think that it may have originally come from space, perhaps in an asteroidal impact, but that it has had billions of years to form and accommodate itself to conditions on Mercury. Evolution seems to be prominent in our Solar System. Who knows what we may find in other star systems!

Anyway, give my best to our friends in Sector P. Remember what I said about the funeral. Actually, I'll dictate and send you some comments, and you can use them as you see fit. Take care.

Love,
Sam

Mercury, Day 57

The temperature continued to rise. On the 57th day, Tilda reported that it was over 350 °C outside and that the electronic shielding system in the hull was being stressed.

"We're in the high green zone right now," she announced during a special afternoon briefing. "The computer indicates that we should anticipate another 70 or more degree increase over the next 10 days. Alone, that wouldn't be a problem, but added to this is the effect of the signal being transmitted by the network—it's higher than our computer models predicted. Together, the heat and radiation from the Sun and the network could tax our shielding system and put its activity into the red zone."

"Can we divert more power to the shields?" Samantha asked.

"It's not a matter of power availability—we have a fission reactor in the engine that can produce tons of power. It's a matter of the shielding electronics themselves and the amount of radiation energy they can deal with. Right now, the system is able to block or deflect most of the radiation away from us, although it is unfortunately least effective for ELF waves. When the system achieves maximum efficiency, then we are left with just the metal alloys in the hull to protect us. That will work for a time, but then the heat and radiation will penetrate into the interior of the ship, threatening the internal electronics, especially the computers in the operational and life support systems."

"And us!" Lars remarked.

"We may have to launch soon," the Captain announced. "Anthony, run some scenarios about how quickly we can launch given different heat and radiation figures. I want to be ready for any contingency."

"Yes, Sir," the pilot responded.

"Chuck, increase your inspection of the ship's electronic systems, including the computers, to make sure they are in tip-top shape. And keep in close contact with Tilda about the radiation."

"OK, Captain."

"Lars, make sure that we are all healthy and sound and that the radiation is not affecting us physically...or mentally. Samantha can help you with this."

They both nodded.

"I will get on the comm for further instructions from Luna City about what we should do as soon as I hear from Anthony."

"Sir, if we do leave, what do we say about the network?" Samantha asked.

"We tell the truth. And if this thing begins to threaten us, we have options of dealing with it."

"What do you mean?"

"I mean that it may have to be neutralized in some way."

With that, he proceeded to the elevator to go up to the fore-deck to contact his superiors in Luna City.

Luna City, Day 58

Admiral Harvey placed a conference call to Army General Suzuki and Marine General Santini. It had top secret security protection.

"Gentlemen," he began. "I want to give you both an update on where we are with the Mercury situation. To review, the *Hermes* crew has discovered an alien presence on the planet that seems to be the source of the signal. It's some kind of crystal network that can generate extra low frequency waves when it gets heated up by the Sun and some underground molten magma, or when it gets excited from some outside stimulation. We know that these waves can be detected from space and that they likely were responsible for the destruction of our scout ship when it dove down close to the surface to investigate the source of the signal it was tracking."

"Yesterday," he continued, "Captain Kilborn briefed me on the current situation. He told me that the network has been transmitting a high intensity EM signal dominated by ELF waves that, together with the rays from the Sun, are challenging the *Hermes'* ability to deflect the resulting heat and radiation. He emphasized that the radiation seems to be some kind of byproduct of the network's normal physical reactions to the daytime conditions on Mercury. He asked for clearance to abort the mission and return home should conditions deteriorate further, and of course I agreed. I don't anticipate any disagreement from the Space Council over this approval."

"However, I think that the destructive potential of the network has been demonstrated by what it did to our scout ship and to the two deceased *Hermes* crewmembers. We need to be mindful of this in terms of the safety of the remaining crewmembers and in terms of any future contact we may have with this entity. It could surprise us at any time with another sudden and destructive action. I think we may need to be more proactive."

He paused for comments.

"I concur," General Suzuki responded. "What are you proposing we do at this point?"

"As you know, the Space Navy has four fusion-class battlecruisers near Venus protecting the security of the orbiting space stations that are located there. My proposal is to detach one of these ships, the *Invincible*, under the command of Captain Martha Granger, and to have it proceed to Mercury. Given the relative distance between the two planets with respect to their orbits around the Sun, the *Invincible* should reach the vicinity of Mercury in a little over a week from now."

"And what would be its orders?" Santini asked.

"I would direct Captain Granger to simply orbit the planet and await future developments."

"What if the network becomes a threat?" Suzuki asked.

"Our first action would be to encourage the *Hermes* to evacuate. Captain Kilborn already has been given the authority to do this on his own should he feel it becomes necessary."

"And if this is not possible for some reason?"

"Then we would immediately take steps to destroy the network, especially if the *Hermes* crew is under attack."

"And how would we do that?" asked Santini. "Would you send down some of my marines from the *Invincible*?"

"Probably not initially. Whatever might threaten the *Hermes* would likely be dangerous for ground troops as well. Plus, we know that a space ship flying close to the surface can be destroyed by the network's ELF waves, as witnessed by our scout ship. This would not bode well for a descending shuttle filled with marines."

"Then what?" asked Suzuki.

"I've discussed possible scenarios with my staff. Our belief is that high intensity laser bombardment from space would likely not be effective. This thing seems to feed on the light energy from the Sun, and the lasers might simply make it more powerful. Blasting it with large fusion bombs might destabilize the surface and threaten the *Hermes*, which is only 120 km away. The best option would be to send down high-speed missiles loaded with small fusion warheads to destroy the network."

"Would they successfully reach it?" asked Suzuki?

"Probably, but we don't know for sure. The missiles move pretty fast and have a computer-controlled avoidance system. They certainly have more diversion capability that our scout ship had, which was moving slowly in a straight line while it took optical images of the surface. Our best guess is that one of the missiles would get through to the target."

"Any other options?" Suzuki persisted.

"Not at this time, General."

"Then the Army approves of your plan. It seems like the best contingency available to deal with this situation."

"I also agree with this course of action," added Santini.

"Good, then we'll proceed with this plan. Secrecy is in order here—we don't want to alarm anyone, especially the public, and especially if the missile option does not need to be executed. Hopefully, we won't have to resort to military force. But if we do, then we'll be ready. Thank you, gentlemen."

With that, Harvey signed off.

Mercury, Day 62

As the outside temperature continued to rise, the network became more active. Light energy was focused to provide more heat energy, which was transmitted down below. To this was added the heat from the magma in the volcanic vents. More molten silicon formed from the silicate rocks in the area. Silicon ions churned and circulated even faster. As a result, the intensity of the ELF waves increased, which were radiated across the surface of Caloris. And the network became even more alert and aware.

"Hail to Maker," it thought. "More heat from above. More growth down below."

It was content, for things were as they should be. It was really awake now. It could rejoice and bathe in the light of the Maker.

Mercury, Day 63

As they returned to the *Hermes* from their latest excursion, Chuck felt unexpectedly exhausted. He had experienced a nagging headache all day, and he was looking forward to a shower and some rest. He and Samantha had placed six radiation detectors 4 km from the ship that stretched out into a line 1 km long. All were pointed in the direction of the network. On their way back, they had done some rock collecting.

"I'm going to get these samples into storage and then take a nap," Chuck sighed as he parked and secured *Rover 2*. "I really feel wiped out."

"Me too," Samantha said. "I guess our activities were more strenuous than we anticipated."

Chuck turned toward Samantha.

"You know, these excursions close to the ship seem like so much busy work. It's really a shame that we can't go back to the network due to the high radiation levels. It's been over a week since our last visit. And the Captain has continued to refuse an expedition to one of the poles."

"I know," she sighed. "I'm not sure what we're accomplishing scientifically. I think we're close to aborting the mission, so the Captain wants us all close at hand."

After they entered the ship and took off their space suits, they went up to the mid-deck. There, they were greeted by the rest of the crew.

"Welcome back," Kilborn said.

"Thanks, Captain, it's good to be back," Chuck said. "All of the detectors are in place and seem to be working properly."

"Yes, they will give us even more sensitivity in monitoring the radiation," Kilborn said.

"What's been happening with the level?" Samantha asked.

"Not good news. Tilda tells me that the radiation level has continued to go up, and the radiation shields are almost in the red. But the doc is concerned about you."

"Damn right I am," Lars said, walking over to them. "How are you both feeling?"

"A little fatigued, really drained," Samantha said, looking at the concern in Lars' eyes. "We had thought that maybe the high temperature outside was overloading our life support systems."

"The outside temperature may be playing a factor. . .it's over 380° now. But the ELF radiation level is 5 % higher than yesterday. It seems to be accelerating in its intensity."

"Wow, that's getting up there," Samantha said. "What's the plan?"

"We'll track the radiation with the new detectors you set up for a day or two," Kilborn said. "But if this rate of increase continues, we'll have problems. Our electronics and computer programs will go bonkers. We will have to evacuate."

"Anyway, you two get over to the exam table," Lars said, pushing Chuck and Samantha away from the rest of the crew. "I want to take a look at both of you. The ship is doing fine right now."

They went over to medical section. While Chuck entered their excursion report on the central computer, Lars began his examination of Samantha.

"Sam, this whole business worries me," he said quietly. "It seems like we're sitting ducks, just waiting for things to get worse."

"I know what you mean," she responded. "We're prevented from making an excursion out to the network, but that's the best way we have to understand it. Maybe there's another way. Maybe we can communicate with it."

"How?" he responded, putting down his stethoscope and activating the vital signs indicator.

"I'm not sure yet, but we have to find a way to tell it that we're not a threat. We have to get it to slow down its metabolic processes somehow so that there isn't so much radiation."

"That would be a good thing," Lars said, recording her temperature and heart rate in the central computer.

"But we know the Captain has contacted his superiors in Luna City. I'm worried that with the military involved, things could escalate in some way that might destroy our chances of better understanding the network. We might be ordered to evacuate. Or maybe they would decide to try and destroy it."

"Why would the Space Services do that?" Lars asked as he tested her knee jerk reflex.

"I don't know—perhaps because they see it as a threat, or perhaps for safety or security reasons."

"Maybe you're right about your suspicions. The Captain has certainly kept his military communications a secret." He pulled out the portable EKG machine and put the single omniprobe over her chest.

"And Anthony as well. He knew all along about the destroyed scout ship."

"Yes," said Lars, "but he also has been under military orders. He couldn't undercut the Captain's authority."

"You're right, but I find this whole business troubling."

"For sure," Lars said.

After concluding that her heart was functioning normally, he detached the omniprobe and motioned to Chuck to have his turn on the table.

"I guess we just need to be vigilant."

"I agree," Samantha said. "Well, how do I look?"

"You look great! And your physical status is normal as well."

They both laughed, but both also were mindful that the future was unclear and potentially dangerous.

Mercury, Day 67

The temperature below increased to a critical level where nearly all of the available silicates became fluid, ionized, and dramatically increased in motion. The result was a surge in ELF wave production. It rejoiced.

"Much growth will take place. Maker has blessed me."

The radiation surge projected in all directions. It was so strong, the ELF waves even penetrated and went through rocks. Toward the east, the waves that encountered the Hermes finally ceased to be diverted by the electronic shielding system and began to penetrate the metal itself.

This meant nothing to the network. It simply thanked the Maker for its bounty.

Mercury, Day 68

In the early morning hours of the 68th day, an alarm went off! Tilda, who had been on night watch in the fore-deck for just such an occasion, jumped out of her portable cot and rushed over to the control panel, which lit up like a Christmas tree. She checked the central computer and glumly looked at the rest of the crew who had rushed up from their sleep pods in the aft-deck.

"We've just received a surge of EM energy. It penetrated the ship's radiation shields and began to interfere with some of the computer programs. The sensors are way in the red now. I'll need some time to fully assess the damage."

"OK," Kilborn said, rubbing his eyes. "See what you can do."

They all moved to their emergency duty stations. Chuck went over to help Tilda investigate the integrity of the computers and life support systems. Kilborn, worrying about the safety of the mission, consulted with Anthony, and the two of them began developing plans for a launch. Lars and Samantha recorded the radiation and thermal levels in the ship and discussed possible biological ramifications of the surge.

After a while, Kilborn called an emergency meeting. They all gathered except for Chuck, who kept examining the computers.

"Well, Anthony and I have some bad news. We didn't expect such a sudden surge of activity. The damage is so extensive that we can't launch right now. Chuck is trying to get the computers back on line. We can't even communicate with Luna City."

"So, what do we do?" Lars asked. "The interior radiation and thermal levels are higher than normal, but they aren't biologically lethal. . .yet. The metal in the ship's hull is giving us some protection, but not for too much longer."

"How much time do we have before the computers shut down completely?" Kilborn asked Chuck, who was across the room.

"I don't know," he responded. "Maybe at this rate, a couple of days. Can we figure out a way to launch by then?"

"Anthony?" the Captain said.

"It's hard to say. I can get us off this rock within 90 min, once we get the computers and other electronics fully operational. That's the main limitation right now."

"It is indeed," Chuck said. "The central computer is giving me some strange messages. I can try to reprogram it, but we've got to keep the ELF wave levels down. If they continue to increase, this will not bode well for the electronics."

"Keep at it, Chuck. Tilda, can we do anything about the radiation?"

"No, Sir, the shielding is at its maximum efficiency, but the ELF wave intensity is just too high for the system to handle."

"All right. Keep trying to reach Luna City."

He turned to the others. "If anyone has any bright ideas, I sure would like to know them."

Luna City, Day 68

Admiral Harvey grumbled as he was called away from his staff meeting to the Red Phone in the War Room. Protected by a series of shields and electronic scrambling devices, this special phone was about as secretive as anything could be in 2130. On the hologram being projected above the phone was an image of John Dexter, the Administrator of the Space Council. He looked worried.

"Hello, John," Harvey said.

"Hello, Admiral. I'm sorry to interrupt your meeting, but we've lost contact with the *Hermes*, and I thought you'd like to know right away.

Damn right, you moron, the Admiral thought, but he said: "Yes, Administrator, I certainly am glad you called. What happened?"

"Earlier we were receiving a report from Tilda Chang that the ELF waves from the network had increased, when her message became static and then cut off. We suspect that the waves had done something to interfere with the transmission."

"Yes, that's certainly possible," Harvey said, thinking about the same occurrence that had affected the scout ship before it went down.

"We're doing what we can to reestablish contact," Dexter continued.

"Is there anything you want me to do at this point?" Harvey asked.

"No, just hold on for now. I'll get back to you as soon as the situation changes."

****

As soon as he hung up, Harvey made his own call to the Space Navy's Deep Space Communication Center. A sleepy-looking Lieutenant, Junior Grade, appeared in the hologram, who immediately brightened up when he saw who was calling.

"Lieutenant, I want you to put a call in for me immediately to the Captain of the Battlecruiser *Invincible*. A private one-to-one link, security level 4."

"Aye, aye, Sir," he responded.

Two minutes later, the holoimage of a middle-aged woman appeared. Her short hair was dyed jet black, and her green eyes stared directly at the screen. Her perfectly cut uniform hugged her trim figure, and it displayed several battle ribbons, including one for bravery during the Mars Insurrection.

"Captain Granger, here," she said efficiently. "How can I help you, Admiral?"

"Captain, we just got word that the *Hermes* has gone silent—there's no comm link. We think that it's due to a sudden and rapid escalation of ELF waves penetrating the ship and interfering with its communication ability. Where are you now with reference to the planet?"

"Sir, we've achieved Mercury space and have been orbiting the planet awaiting further orders."

"Good. I want you to go on standby and prepare to initiate Operation Candlesnuffer. I will contact you later when we're ready to initiate action."

"Aye, aye, Sir," she responded tersely. "Is there anything else, Sir?"

"No, not for now."

"Thank you, Sir," she said, and the screen went blank.

Mercury, Day 68

"Captain, I've just reestablished contact with Luna City," Tilda announced just before lunch. "I can't say for sure how long it will last, however."

"Good," Kilborn said.

He ran over to the comm and quickly gave his report. In the time it took for the signal to reach its destination, the comm line went dead.

"Sorry, Sir, we've lost the link again."

"Keep trying to reestablish it," Kilborn said.

He walked back over to where he and his pilot were analyzing their launch options.

"Well, Anthony, I briefed my superiors at the Space Services Center on our situation. I don't know how much of my report will reach them, but at least they have some idea of where we stand. I have no clue, however, as to what they will do."

"Do you think our 'help' has arrived?"

Kilborn went closer to Anthony and whispered.

"I received a message that the *Invincible* has arrived in Mercury space. I don't know what the plan is for now, but we must be prepared for anything."

The Invincible, Day 68

Captain Granger walked briskly toward the Control Room. She just heard from Admiral Harvey on a private comm line that the Space Council had been briefed and that Operation Candlesnuffer had been approved and was in force. She entered the room, made her way to the Captain's Chair, and activated the ship's intercom.

"All hands, listen up!" she announced, her voice booming throughout the *Invincible*. "We are to proceed to the air space above the network and destroy it immediately. Navigation, get me over the target! Ordinance, prepare four fusion light missiles for launch! Computer, initiate all ship defense systems and activate full visual screens! Communications Officer, open all channels and report any messages! I want a continuous radio relay broadcasting our activities back to Luna City!"

There was a flurry of activity as the lethal ship began to come to life.

****

In its hyperalert state, it became aware of messages being transmitted from above. Something was flying overhead, then stopping directly above it. This was like the transmission that had happened before, when it needed to destroy an intruder.

"Another visitor from above. Will this come down and destroy part of me? Is it a danger?"

Agitated, the network began to respond. Waves of varying frequencies danced through its system, especially ELF waves, as it began to experience a fear response.

****

"Captain, we are over the target," came the message from Navigation.

"Good. Ordinance, are you ready to launch?"

"Yes, Ma'am," came the response.

"Then launch the missiles at will!"

A shudder was felt in the Control Room, then a second one, a third, and finally a fourth.

"Missiles launched," came the response as the deadly rockets flashed downward.

****

It was aware that several weaker transmissions had dissociated from the larger one. Four objects were rapidly descending toward it. It sent a bolus of ELF waves skyward in a broad rising umbrella, headed for the intruders.

"Must destroy," it thought. "Must survive."

****

"Captain, there has been a surge of ELF activity coming from the target," announced the Communications Officer. "It's rapidly expanding into a broad wave front. The missiles are starting to initiate diversion action, but the front is too wide and moving too fast for them to escape."

In visible light, the monitors showed each of the rockets exploding.

"Damn!" Granger said.

"Captain," the Communications Officer continued, "the front is moving rapidly toward us. It's of high intensity and broadening even more! Captain, it's almost here!"

"Activate emergency evacuation," Granger announced. "Let's get the hell out of. . ."

****

Monitoring her screens just before dinner, Tilda Chang was astounded by what she was seeing. So enraptured, in fact, that she neglected to inform anyone else. Four bright pinpoint lights hurled down from the heavens from a larger object that was centered directly over the network. Then, a massive ELF wave front ascended skyward, leading to the explosion of the four smaller lights. A moment later, a larger explosion erupted above the smaller ones. What happened? She sent a notification and a holoreplay of the event to Captain Kilborn, who was in his quarters at the time.

Luna City, Day 69

A somber Admiral Harvey appeared before the assembled group waiting for him in the Space Services Conference Room. Besides the usual military council members, Administrator Dexter and members of his Space Council had also been invited to attend, along with representatives from the Solar System news media. This emergency news conference had been called to mitigate the rumors that had gone viral through the social media about the destruction of a battlecruiser near Mercury and a possible alien invasion of Venus and other planets in the Solar System.

"Ladies and gentlemen, I have bad news to report," Harvey announced. "As most of you know, the Space Services activated a plan to send the Battlecruiser *Invincible* on a military action to Mercury in an attempt to save the crew of the *Hermes* from a hostile action. The *Invincible* entered Mercury space and was ordered to attack an alien presence on that planet, a network of silicon crystals and fibers that is able to concentrate energy from the Sun and from the interior of the planet. The radiation from this energy was threatening the crewmembers on board the *Hermes*. The *Invincible* launched four fusion light missiles at its target, but they were destroyed by a wave of energy emitted from the alien. Furthermore, this wave front expanded and was able to engage and destroy the *Invincible* itself. I am sorry to report that all of the crewmembers were killed during their brave action. Relatives have been informed, and a memorial service is being planned."

There was a murmur from the audience.

"Are there any questions?"

A hand went up.

"Yes," said Harvey, "the gentleman from the Mars City news consortium."

"Admiral, how could this happen to one of our newest and largest Space Navy fusion ships? The briefing report that you sent us just before this meeting said that this...network...is only 50 m across."

"That's true," he responded, "but apparently it's able to harness a vast amount of extra low frequency, or ELF, wave energy of sufficient intensity to travel into space and penetrate our radiation shields. It can dismantle electronic systems and generate enough heat to be lethal to both non-living and living systems."

"By the latter you mean the crew of the *Invincible*," stated a woman standing next to a camera labelled *Asteroid News*. "But weren't you aware of this possibility? We've heard rumors that an earlier naval scout ship was also destroyed by the network while flying over it."

"There's always a risk in combat," he answered, ignoring her last comment. "Our battlecruiser was high up in space, and the speedy missiles it launched had a diversion system. We didn't expect that the alien would produce such a broad wave front of such high intensity."

"Is this a danger to our orbiting satellites around Venus, or anywhere else in the Solar System?" asked a reporter from the back row.

"Probably not. We don't think that the alien is able to project its ELF energy beyond Mercury and its environs."

"What about the *Hermes* crew?"

"We're still working that situation. The alien has been generating ELF waves as part of its...normal functioning. These have been leaking out and encountering the *Hermes*. The shields have already been breached, and the damage to the on-board computer and electronic systems has compromised the ability of the crew to launch their vehicle. Furthermore, the waves have been increasing in intensity, so the situation is becoming ominous, unless we do something."

"What do you have in mind, Admiral?" asked the Mars City newscaster.

"We've been exploring the possibility of another attack from space, but we need to first devise a way of protecting ourselves from a counterattack. Our best weapons analysts are working on the problem."

"Is there anything that the *Hermes* crew can do?" asked a *New York Times Holovision* correspondent. "I understand that they've been able to get close to the network and even interact with it."

"That's not possible right now. The radiation levels are too high."

"What about a response from the *Hermes* itself. We've heard rumors of a defense system located on-board."

"I can't comment on that at this time."

"But Admiral, perhaps they could do something to help themselves. Perhaps it would be possible...."

"Again, I can't comment," Harvey interrupted. "We have the best minds available analyzing the situation."

He paused.

"Well, that's all we have time for now. Thank you for your attention."

He turned and walked stiffly out of the room.

Mercury, Day 69

Captain Kilborn assembled his crew in the mid-deck for a special meeting in the afternoon of Day 69. He walked somberly up to the dining table, around which the crewmembers were gathered.

"We had a comm window open up this morning and were able to receive a message from Luna City about what happened yesterday," Kilborn said. "The message was from the Space Council. It said that a Space Navy battlecruiser attacked the network in order to try and save us. Unfortunately, the network won the battle. It unleased a massive ELF wave front that was directed upward toward four missiles launched by the battlecruiser. The four small explosions that we picked up were the missiles being destroyed. Unfortunately, the large subsequent explosion was the battlecruiser itself, with all hands lost."

There was a stunned silence. Samantha spoke first.

"Captain, I'm truly sorry about the loss of lives. . .all those men and women. Their attempt to help us was truly heroic."

"Yes," said Anthony, "there are hundreds of people on a battlecruiser. It's a fusion-powered ship with the latest technology. All gone—in a puff of smoke!"

"So what happens now, Captain?" asked Chuck. "Will there be another response from the Space Navy?"

"The message didn't say. I'm sure the Space Services will be considering a number of scenarios at this point in time. Tilda, what's the latest on the ELF radiation?"

"It continues to penetrate the hull. We're doing what we can to maintain the temperature by diverting energy into the cooling systems, but soon even they will be overloaded."

"Chuck, how are the other systems?"

"Not good, Captain. I continue to reprogram our computers, but they're running slowly, with glitches appearing all too often. Life support systems are still operable, but who knows for how long? As you know, our comm line comes and goes."

"How about outside?"

"The outside temperature is just shy of 400°, but it will rise even more. What's worse, the Sun won't set for another 19 days, and the temperature will remain high for a while afterwards."

"OK, thank you. Anthony, I think it's time for us to take action and activate Plan Omega."

"Aye, aye, Sir."

The other crewmembers looked at each other, puzzled.

"Sir, we don't understand," Lars commented.

"When we were planning the mission, there was always the possibility that the source of the signal would be dangerous and that our lives would be at risk. Indeed, that appears to be what has happened. Evacuation was always the first option for us, but that's not a possibility right now. We hoped that the Space Services would come up with a plan to rescue us, but as we have seen, their attempt to help didn't work. The last resort was Plan Omega: we destroy the source of the danger ourselves."

"How," asked Samantha.

"With a small nuclear fission missile."

"What! How?" asked Lars.

"Embedded in a sealed storage compartment near the rear of the ship is a ground to air missile with a small nuclear warhead. It has its own power supply and guidance system. It takes off, fixes on the target, and blows it up. The presence of this missile was never revealed to the press, nor to anyone else except for the top brass in the Space Services, me, and Anthony."

"Where is it?" asked Chuck.

"In one of the four rear supplementary supply holds," Kilborn said. "It was manifested as parts for a space telescope that we were authorized to launch on our way home, but that was a ruse."

"What are in the other holds: fusion bombs?" Chuck asked.

Ignoring the sarcasm, Kilborn responded: "No, extra food, water, oxygen, an emergency nuclear power generator—the material legitimately listed in the manifest."

"Captain, will the missile be operational?" asked Tilda, who also seemed surprised at this revelation. "With all the radiation interfering with our systems. . ."

"Right now, yes," interrupted Kilborn. "Anthony just completed a systems check. The missile has its own independent heat and radiation shield system, which should protect it in the brief time it will take to reach the network. We plan to launch it as soon as Anthony opens up the outer hatch to the hold, takes it out, and activates it."

"But Captain," said Lars, "the network is likely to detect it as it arcs into the air and destroy it, similar to what happened with the battlecruiser missiles."

"Our device can be programmed to travel at supersonic speeds close to the surface of Caloris—no higher than 5 m. Being in the middle of a high plateau, I believe the network will not be able to detect the missile until it comes over the edge, when it will be too late."

"We can't destroy the network!" Samantha blurted out. "It may be conscious. We've never found anything like it before. Think of it—a silicon-based life form that has awareness and emotional responses to its environment!"

"It's a dangerous life form, dangerous to us and to anyone else who gets close to Mercury," Kilborn responded. "No, Samantha, it must be terminated."

"But it..."

"That's enough!" he growled. "I have my orders, and around here, my orders are the law. We will destroy this thing."
****

Anthony and Chuck suited up and went outside. They accessed the missile and its launch system. Anthony programmed it with the network's coordinates and manually set the launch sequence. The crewmembers waited in the foredeck for their return. Anthony and Chuck returned to the airlock shortly after 1600 h, took off their suits, and called the Captain.

"All set, Captain," Chuck announced somberly over the airlock intercom.

"Nice job," Kilborn said.

"Thank you, Sir. The systems responded well, and the programming went just as planned. The missile is good to go. We're on our way up to the foredeck."

"See you soon," Kilborn said.

He then pushed some buttons on a portable launch console.

"I hope this works," he said.

A buzzing sound was emitted. Then, after a short pause, a slight vibration was felt in the ship, then quiet.

"It's off. Now we wait. It should get there in a little over 4 min," Kilborn said.

Anthony and Chuck joined the rest of the crew, and they all gathered before the monitor tracking the trajectory of the missile. The waiting time seemed like an eternity. No one spoke. As Samantha looked around, she noticed that her crewmates seemed tense, apprehensive. They appeared to be hopeful for success. Why, then, didn't she feel that way?

Part of me wants to get rid of this source of danger, she thought to herself, *so that we can all live. But part of me wants the network to live, too. It's not trying to hurt us. It's just trying to survive, like any life form. We are the intruders here, so we are the ones who should die.*

The upper right quadrant of the monitor showed an image produced by a camera embedded in the front of the missile. It showed the surface of Caloris racing speedily by, and up ahead the targeted hill came into view. It quickly increased in size. As it was approached, the missile's speed slowed, and the side of the hill dropped as it ascended. The rim was quickly reached as the missile crossed over the edge and flew onto the plateau.

****

It sensed a wave front coming from the direction in which the Maker rose during every awake cycle. It became alert and fearful, causing its ELF waves to increase in intensity and to crisscross rapidly through its rods and fibers.

"Something unknown is happening. It must be danger."

It focused its thought energy in the direction of the oncoming radio waves. Suddenly, they stopped, and just as suddenly appeared again with great intensity, over the edge of the plateau.

In a microsecond, the network discharged a bolus of ELF waves. Something disintegrated, with the debris crashing onto the plateau, creating numerous small craters and sending rocks and dust into the air. Then, all was quiet.

****

"Whoa," said Tilda. "I just recorded a brief burst of ELF waves, then a ground tremor coming from the direction of the network."

"I didn't see a nuclear explosion," Kilborn said, looking intensely at the monitor. "Was there an explosion?"

"Not that our instruments recorded," Tilda said. "It appears that the missile disintegrated and crashed on the plateau, but no nuclear devise was triggered."

"The damn thing must have fried the warhead electronics before they could detonate," Chuck said. "This must have happened as the missile cleared the rim of the plateau. We're talking about a very brief period of time."

Tilda and Chuck shook their heads. Lars looked stunned. Anthony put his head down on the table. The Captain, still holding the launch console, showed no emotion. Samantha sighed, suppressing a smile.

****

"Danger gone," it thought. "Intruder destroyed. But it did not come from above. Where did it come from?"

It tracked the radio signature back toward the east. It followed this track with a signal of its own until something interfered with it. In a process similar to echolocation in whales, it perceived the characteristics of the object by the way it reflected the signal. The object did not bounce it back like a naturally elevated object on the surface. Instead, the reflection pattern resembled the one that it had received from the intruders that attacked it from above. It also resembled the reflection pattern from the visitors that had come to it earlier, then left.

"But object is far away. It must be very large. Maybe it is Maker of the visitors."

It became confused. Part of it wanted to destroy the elevated object. Part of it wanted to understand it better. Was it a friend or an enemy? Would it be a danger again?

Another idea entered its consciousness.

"Because it exists, I am not alone."

Mercury, Day 70

The outside temperature exceeded 400° on Day 70. Originally, this milestone was supposed to trigger a celebration, but no one on board felt like celebrating. The inside temperature would sometimes spike to over 40° as the cooling system struggled to keep up with the heat. Every time Chuck repaired the electronics and reprogrammed the central computer, the system again would go haywire due to the penetration of the ever-increasing temperature and radiation.

That afternoon, Lars came over to Kilborn, who despite the heat was exercising on the treadmill. The doctor's eyes were bloodshot due to lack of sleep from the night before.

"Captain, the lab chemistries are showing all kinds of stress responses in the crew. In addition, I'm getting irregularities in our life support system, despite all the built-in safeguards. The carbon dioxide scrub has been continuously turning off and on, and the coolant regulator is not working at nominal efficiency."

Head throbbing, Kilborn glanced at Samantha, who had joined them. She looked glum and washed out.

"The heat and radiation are affecting all of us directly," Lars continued. "Our blood pressures and pulses have been jumping around, and the head-aches have made it hard to think clearly."

"Well, I'm not through yet!" blurted the Captain, wiping sweat from his brow. "This alien thing is not going to get the best of me."

"Who's the alien?" Samantha retorted. "This...alien...has been here a long time. We've been here less than 3 months."

"Well, whatever it is, it's not doing us any favors. We'll fry before too long. Being barbecued on a charcoal planet isn't the way I planned to check out."

They both watched as Kilborn jumped off the treadmill and stomped over to the mid-deck computer console to record his exercise numbers.

Lars turned to Samantha.

"Any bright ideas about how to get out of this mess?"

"Not really. I'm planning to take another look at the crystal fiber samples under the microscope. Possibly, I'll get some insight about how the network generates its radiation."

"It's worth a shot, Sam. Maybe if I go back over the results of our experiments on the plateau and think more about the network as a life form, like us, I'll get some ideas as well. Perhaps we'll think of some way to communicate with it and tell it to slow down its metabolism and stop generating so much radiation!"

She smiled weakly.

"Anyway," Lars continued, "one way or another, we'll get out of this."

They walked over to the lab.

As Samantha seated herself on the stool behind one of the microscopes, she wished that she could feel more optimistic about what Lars had just said.
****

That night, Chuck was able to temporarily repair their comm system. Waiting for Samantha was the latest message from her mother. With great excitement, she went to her sleep pod and pulled it up on the computer at her desk.
****

To: samantha.evans@Mercuryexplore.assc (RESTRICTED USE)
From: k.evans@stationdelta.org
Subject: Hello
Date: Mercuryexplore, Day 68

Dearest Samantha,

I don't know when this message will reach you. . .maybe in a couple of days. I understand that the comm links to and from the *Hermes* have been off and on. Anyway, the media have been flooding us with the news that a battlecruiser that tried to rescue you failed, and that all the people on board were killed during the attempt. My heart goes out to their families and friends—what a tragedy!

It appears that the alien you've encountered is very powerful and very malevolent—at least that's what they're telling us here. They say that you're trapped on Mercury and that this so-called network is generating energy waves that continue to interfere with your equipment and is causing some crew distress. I hope that all of you are OK. I worry so much about you and your crewmates. I pray that you will be able to launch and escape the clutches of this thing. Perhaps there will be a lull in the radiation so that you can get back home safely.

The funeral for your father went very well. Several people spoke favorably of him: fellow workers, long-term friends, even nursing facility staff. And your thoughts were represented: John Gomez read the eulogy you e-mailed to me, and it was lovely. John had a bouquet of flowers sent over from the hydroponic gardens. You remember how much your father liked flowers and how he would sometimes surprise us with a rose or two when he came home. John's thoughtfulness was very sweet. After the funeral, Dad's name was entered in the station's historical log. I've set up the memorial plaque in our bedroom. I feel sad at times, but also a sense of relief that he led a good life and was content and happy most of the time.

My job is going well, and everyone at work sends you their best.

We received another shipload of immigrants, and we're trying to make room for them. We're becoming a bit overcrowded. The news from Earth is that they continue to build more desalination plants to get drinkable water and new fusion reactors to get more clean energy. I understand that the high-tech companies in San Francisco and Austin are leading the way in making new technological breakthroughs. It just goes to show you what humans can do if given the right motivation and resources. Hopefully, the Great Pollution will stabilize, and Earth's atmosphere will start to improve.

I guess that's all for now. Send me a message when you can to tell me how you're doing. I miss you.

Love,
Mom
****

To: k.evans@stationdelta.org (RESTRICTED USE)
From: samantha.evans@Mercuryexplore.assc
Subject: Hello
Date: Mercuryexplore, Day 70

Mom,

Your last e-mail arrived, but 2 days late. I'm sending you this message in response, with hopes that it gets through sooner rather than later.

We're doing pretty well here, although as you know the radiation is interfering with our equipment and electronics. It was indeed sad about the battlecruiser. But Mom, the network is really not malevolent, at least in its intentions. Whenever something stresses it, or threatens it, it has a kind of fight-or-flight reaction, except that it can't move away. Instead, it increases the intensity of the extra low frequency waves that it produces. It's like when you're stressed, you become hyperalert, and your heart rate goes up. The waves help it to deal with a threatening object.

These ELF waves also seem to be related to how the network reacts to the light from the Sun. They are part of its functioning, its metabolism. We think that it breaks up silicon compounds into pure silicon during the heat of the day here on Mercury, then this silicon crystalizes and is added to its body during the local night, when things cool down. The moving silicon ions create the radiation. Lars and I also think that the waves are analogous to our brain waves, and that the network is conscious and can think and feel. Unfortunately, all this radiation is not good for us or our equipment. But I'm hopeful that we'll figure out a way to launch and escape our situation.

Too bad about the overcrowding on the station. Earth really messed up over the past two centuries, and it's unfortunate that the politicians and other people didn't take the global warming threats more seriously. As a result, most of the activities on Earth have needed to be diverted into dealing with the climate changes instead of advancing science and making life rewarding for everyone. But I'm kvetching! At least they've gotten some control over the situation, and perhaps in time Earth will become more habitable.

I'm glad that things went well at the funeral. I know that Dad was well-liked, and he was a terrific father. I still miss him terribly, as I'm sure you do, too. But the tears have stopped now, and my grief is ebbing away, little by little. When you think about it, send me a holoimage of the memorial plaque—I would like to set it up in my room.

I will try to send another message in a few days. Take care.

Love,

Sam

****

After completing her message, Samantha got ready for bed. She reflected on what her mother had said about the funeral and on the comments she sent concerning her father. She reminisced about the happy times her family had when she was a girl. She was glad that her father had some pleasant experiences in the nursing facility. It was interesting how during his music therapy he was able to respond to the sounds and the lights in ways that words could not adequately reach.

Communicating with someone is so important, in whatever way works, she thought.

As she climbed into bed, she was reminded of what Lars had said earlier about talking with the network. That got her thinking about their current situation as she drifted off to sleep. . .

Mercury, Day 71

The next morning, after the crewmembers had selected their breakfasts from the foodbot and were sitting at the table, Samantha looked up at Lars. He was eating his algae bacon and tofu eggs across from her.

"Lars, last night I was thinking about what you had said earlier about communicating with the network. An idea was in my head as I awoke this morning. Do you think the network is aware of us, that it might be able to consider us as being another life form?"

"Well... it's very responsive to stimulation," he responded. "And you've been picking up some of its emotive reactions. So possibly it sees us as different from itself and as some kind of living entity."

"If that's the case, maybe we can in fact communicate with it."

Kilborn looked over at her.

"What are you getting at, Samantha?"

"Captain, before he died, my father couldn't communicate very well with anyone due to his memory problems, and it was hard to interact with him by just talking. When the nursing facility staff gave him music therapy, with varying sounds and light frequencies, he became responsive. My mother said he even was able to dance with her, to relate to her. Perhaps something like this would work with the network."

"How so?"

"We think that it became aware of the radio signals given off by the space probes and the missiles that were sent to destroy it. And in the experiments Lars and I have performed on the plateau, the network definitely became more activated by the waves we projected to it from the EM generator. So...maybe we could transmit radio waves to it from the *Hermes* in some meaningful pattern, something that would indicate that were are intelligent creatures and mean it no harm."

"How would that help our situation?"

"I think that the network has some control over the radiation it sends out. It seemed to be able to direct the bulk of its ELF waves toward the *Invincible* and toward the missile we sent. Perhaps if it could understand that we are being harmed by its radiation, it would modulate the waves. Maybe we could persuade it to decrease their intensity."

"That's an interesting idea, Captain," Lars said. "If this thing really is aware of its environment, it could tell the difference between random waves and patterned waves produced by intelligent beings. The old SETI efforts used the same idea to search for intelligent life in the universe."

"I suppose it's worth a try," Kilborn said. "Tilda, after you finish your breakfast, let's you and I go up to the fore-deck and send this thing a hailing message and see how it reacts."

"Aye, aye, Sir," she responded.

After she and the Captain finished their meals, the two of them got up and headed for the elevator. They were shortly joined in the fore-deck by the rest of the crew.

****

It became aware of energy being transmitted to it from afar. It was from the direction of the intruder that was sitting on the surface. Was this a danger? Probably not: the energy was constant and weak in intensity. It was not increasing in strength, like the waves from the intruders that had come down from above and damaged it.

It thought of the visitors that had joined it earlier on the plateau. They had sent it some energy, and it had responded. What if it did the same thing now? It began to consolidate and direct its energy toward the distant source.

****

"Captain, the radiation from the network has increased, especially the ELF component," Tilda said from her console, as her crewmates all stood behind her. "Not by much, but it's measurable."

"Well, we got a response," he said. "Our continuous hailing message seems to be getting its attention."

"But it's not conveying any information," Lar said. "It's just a steady broadcast, and the response back is the same thing."

"Yes," Samantha added, "maybe if we changed the frequency and intensity of our transmission, created pulses with a patterned power and time and sequence, the network would catch on that someone is trying to communicate with it. Lars and I got this sort of reaction when we visited the network on the plateau."

Chuck laughed. "Are you proposing to communicate with it via a series of variable pulses, something like the on/off sequence that formed the basis of our old computers. . .or maybe some kind of Morse Code?"

"No, I'm not proposing that," she responded. "At least not yet. I'm simply saying that it may perceive us as trying to communicate with it as one life form to another. That's a good first step. Maybe it will respond in kind."

"It's worth a shot," Kilborn said. "What do we have to lose? Give it a try, Tilda."

"OK, Captain. I'll generate an on/off transmission sequence with a repeating pattern in terms of pulses and intensities."

*****

It became aware that the energy signal from the intruder was changing, sometimes strong, sometimes weak, or even off, but in a repeated pattern. It wondered why it was doing this. Was it also a thinking entity? Was it trying to interact in some way? What if it retransmitted the same pattern that was being sent. . .

*****

"It worked, Captain," Tilda said, monitoring the pulses. "First, the network parroted our initial transmission frequency and on/off pattern, and now that we've dropped the intensity of our signal to a lower level, it's done the same.

It's still transmitting, but at a low enough power that we could activate our electronic shields again."

"Good!" Kilborn said. "Keep your signal the way it is. If this thing's response persists, we can fix things and maybe get the hell out of here!"

"Captain," interrupted Samantha, "if things continue the way they are now, we could pay the network another visit, study how it works, maybe even communicate further with it."

"Samantha, I understand your passion," Kilborn interrupted, "and I agree that this network of yours is an important discovery. But we've been on Mercury longer than planned, and this alien is still a danger. I don't want to risk anyone's life on another excursion to pay it a visit."

Samantha's face reddened.

"But this so-called alien may be alive! Don't you understand? I have to find a way to talk to it. I need to study it more!"

"My orders are to launch and come home as soon as we're able," Kilborn responded with some impatience. "There'll be an opportunity for another crew to come here in a better shielded ship and study your silicon friend more safely. Maybe you can be a member of that crew."

"Maybe they will come back to destroy it," she said. "You tried to do that very thing yourself. Why wouldn't another crew have the same intention?"

"Samantha, that's enough! My orders now are to launch and not risk any lives," Kilborn said.

Her arms tensed stiffly at her side as she lowered her head and turned away, hiding her rage.

"Captain, I agree with Sam, and I have an idea," Lars said.

"Oh brother!" Kilborn exclaimed. "All right, what do you propose?"

"We think the network is responding to our signals. If it keeps the power to manageable levels, why not let Sam and me take a rover and go pay it a visit one more time. We can take our portable EM wave generator with us again and try to communicate with it. Being so close, it might realize that we mean it no harm. We did this before."

"Or it might think that you're hostile and begin transmitting with more power, endangering not only the two of you, but the *Hermes* as well. At any rate, if we wanted to communicate further with it, we could do it by varying the signals we're sending from the ship?"

"Sir," Lars persisted, "if we were on the plateau, we could hook up our modified EEG machine and monitor the network's thought waves. . .or whatever they are. . .like before. If it becomes too agitated, we can back off, and nothing will happen to the *Hermes*. Or, hopefully, to us. Anyway, Sam and I are willing to take the risk."

He looked at Samantha, who had composed herself and was nodding her head.

"We can learn more about it, Captain," Lars continued. "And when we leave, we'll have more knowledge of what we're facing. That will help any returning crews deal with this thing. Including Space Navy crews."

Kiborn laughed.

"What a smooth talker you are, Doctor. You should have been a politician! OK, now that this entity has quieted down a bit, I'll try to contact Luna City. Let's see what they say about your plan. But Anthony and I will set up a contingency plan for a quick launch just in case."

Luna City, Day 72

"They want to visit the alien again?" asked Admiral Harvey. "This thing has already killed two shiploads of naval personnel. Now that the *Hermes* may be able to leave, that should be their top priority."

This time, the meeting was held in the Space Council Assembly Room, with the Space Services military being invited as guests. Administrator Dexter was presiding.

"Yes, Admiral," he said. "That is the plan proposed by Captain Kilborn. It appears that Drs. Ahlgren and Evans are willing to pay the network a visit and attempt to communicate further with it. They're willing to risk their lives. Apparently, the intensity of the signal being broadcast by the network has remained low enough over the past 24 h to enable them to pay it a visit."

"But the low intensity also means that the *Hermes* can launch." Harvey pointed out. "What if they visit this alien and manage to provoke it with their tinkering? What if it decides to fry them, and then the rest of the crew on the *Hermes?*"

"The two doctors feel that they can hook the network up to a modified brain wave machine like in their previous visit and deduce when it's becoming agitated. Should this happen, they would back off on their attempts to communicate further with it and return to their ship. The entire plan is summarized in the briefing report on your personal computers."

"I've read it, Administrator Dexter," Harvey said. "But it still seems risky to me."

"There's always a risk when an alien species is involved, Admiral," voiced Dr. Curry, the Chief of Astrobiology for the Space Council. "Remember the trouble we had landing on Europa and drilling down into the ice until we got through to the subsurface ocean? Not only did we have to follow strict quarantine procedures, but the ice cracks that developed threatened the crewmembers. We were afraid of losing them. But look what happened: we discovered multicellular organic life on a moon of Jupiter!"

"Yes, Dr. Curry, but those little fish weren't sending out death rays!" Harvey countered. "Plus, anytime we send a ship near Mercury, we can't assume that it will be allowed to pass unmolested."

"All the more reason to understand the network even better, perhaps to communicate with it," Dexter said. "The *Hermes'* astrobiologist, Dr. Evans, feels that the network is not malevolent. Yes, it reacts to perceived danger with a hyperalert response, like any other conscious entity. But the network has some ability to modulate the intensity of its waves when it's not being threatened, at least to levels that are safe for the crew. And both Dr. Evans and Dr. Ahlgren believe that they have established a primitive communication with it already, so much so that it has reduced its signal output."

There was a silence. Then, Dexter spoke.

"Well, ladies and gentlemen of the Council, I think the issues under discussion are clear. It appears that we must weigh the scientific advantages of the plan put forward by the *Hermes* scientists with the risk of danger to the crew and to the mission. I suggest it's time to vote on the matter."

Mercury, Day 73

At lunch time on the 73rd day of their stay on Mercury, the Captain came down from the fore-deck.

"I just finished reading a message that came through from the Space Council. They took a vote and decided to OK the plan put forward by Samantha and Lars. The Space Services group was opposed, but since the Council is a civilian agency that has several scientists as members, the vote passed."

Samantha and Lars smiled at the news.

"Tilda, where are we with the radiation?"

"It remains low, Sir, even decreasing slightly. This might be related to the fact that the Sun is descending toward the western horizon, and the outside temperature is dropping as a result. It's currently below 360°."

"Good. That'll cool off our friend a bit. Chuck, how are the shields?"

"I got them up and running. They seem to be diverting the heat and radiation away from the hull for now."

"Anthony, what is your launch strategy?"

"I have a scenario where we could be ready to launch by tomorrow night, Sir. We still have a few more things to repair and some software to reprogram."

"All right, Samantha and Lars, make your plans to take *Rover 2* out tomorrow morning. Neither of you will be needed on the ship to get things ready for launch. If you're willing to be damned fools and risk your lives trying to talk to this thing, be my guest. Just be sure to come back by dinner time. We'll plan to launch after we eat a good meal. When we're ready to take off, we'll do so, with or without you."

"Thank you, Sir," they responded in unison.

****

To: k.evans@stationdelta.org (RESTRICTED USE)
From: samantha.evans@Mercuryexplore.assc
Subject: Hi
Date: Mercuryexplore, Day 73

Mom,

Things have gotten a bit better. In a way, I have you and Dad to thank for it. I remembered your comments that Dad was being treated by music and lights and that he responded to this stimulation better than to verbal communication. So, we tried sending out a radio hailing frequency to the network, and it responded to us. As a result, it lowered the intensity of its ELF waves, and we've been able to repair our electronics and troubleshoot our computers. If this continues, we plan to launch tomorrow night.

Before then, Dr. Ahlgren and I are going to pay a visit to the plateau and try to establish better communications with the network. I don't know how successful we will be, but I just wanted to let you know. I'm optimistic that things will work out. On the other hand, there's a chance that the network will perceive us as some kind of threat. If so, then who knows what will happen. Be assured that Lars and I will be very careful and won't take any chances.

My next message will be in a couple of days. By then, I expect to have some news about the success or lack thereof as regards to our encounter with the network. Or, I may be giving you a progress report concerning our trip back home. I'm not sure which will be the most likely scenario—maybe both things will happen. Anyway, time will tell.

I just want you to know that I love you and am thinking about you. Try not to worry about me and my crewmates. I hope all is well on your end and that you're coping with things. Give my best to John and to the rest of your friends. Talk with you soon.

All my love,

Sam

Mercury, Day 74

The next morning, Tilda reported that the network was continuing to transmit at a relatively low level, and the shields were working adequately enough to protect the ship. Chuck and Anthony continued to prepare the ship for launch. Lars and Samantha met with Kilborn, who gave them permission for their excursion. After they suited up, they rode the elevator down to *Rover 2*. Samantha looked over at Lars.

"So, you're OK to drive?"

"I guess so. I went over the procedures last night. It's a good thing Chuck let me practice during my last excursion with him. What was a fun activity now is a needed skill."

"Just don't run any red lights," Samantha joked.

They entered the rover and took off for the plateau. Initially, their ride was bumpy, especially since Lars managed to find several rocks to go over. But after 15 min or so, his driving skills improved, and the ride became much smoother. As they passed the halfway point, Samantha shut off the comm line to the ship.

"You know, Lars, I don't know what will happen when we encounter the network. But I want you to know how much you have meant to me, both as a friend and as. . .something else. I've valued our time together during this mission."

He laughed and looked at her.

"You sound glum. Be positive. I think we'll figure out a way to talk to this thing and that our launch will be successful. And I think that you and I will have a chance to continue seeing each other, both in secret on the ship and afterwards in Luna City, or wherever."

She laughed. "I doubt our relationship is a secret by now. I've seen Captain Kilborn look at us when we're together and scowl. The same goes for Tilda, but for a more personal reason: I think she has some feelings for you."

"Maybe. But you're the one I care about. There's no way a cranky space ship captain or a jealous comm officer is going to get in our way."

He smiled at her, then looked back quickly as he just avoided driving them into a small ravine. Samantha turned the intercom back on, then did some preliminary checks of their equipment.

As she looked out at the barren landscape and the growing shadows, this time facing east as the Sun flamed low in the western sky, she imagined how hot it still was outside the rover and how their suits would be working to keep them alive when they got out. She marveled at how this bleak environment, scorching during the nearly 3 month daytime and freezing during the equally long nighttime, could harbor a life form. But then, life was adaptable, whether it be carbon- or silicon-based. She wondered what people would find when they finally ventured to the stars.

Interrupting her reverie, Lars announced that the plateau was up ahead. Samantha contacted Tilda on the *Hermes*, who said that the radiation emanating from the network was safe but that they should limit their exposure to an hour or so. They drove up to the slope leading to the top of the plateau and turned off the rover.

"OK," Lars said, "let's gather our stuff and head up."

The pathway was familiar now, and it seemed that they reached the top in no time at all. They bounded over the edge to the network. Lars set up the EEG electrode array around the crystal rod cluster, and Sam activated the EM generator.

"All right," he said, "let's get started."

Sam set the machine to broadcast at standard intensity in the low-frequency radio spectrum, and she pointed the beam at the network.
****

It became aware of another source of radiation, this time close to it. Initially, it was surprised, but then it relaxed as it realized that this type of radiation had occurred at another time. The visitors were back.

"Must not react strongly," it thought. "Must not cause them to go away. Must not be alone again."

It directed its attention to the new radiation source.
****

"I'm getting readings," Lars said. "When you first started broadcasting in the radio range, the network's waves spiked up to around 22 Hz, but they're in the 10 Hz range right now—pretty relaxed."

"Yes," she said, "I felt a twinge of anxiety, but now a sense of peace coming from it."

"OK, drop the frequency a bit. Let's see what happens."

Sam adjusted the generator.

"My readings show a slight drop, to 8 Hz," Lars said.

"All right, now I'll increase the frequency," Samantha responded as she made the necessary adjustments.

"That activated the network, Sam. The EEG reading is up to 13 Hz."

"Now I'll drop the frequency, then put in a rapid sequence of high-low-high-higher-low."

In response to her actions, the network mimicked the changes, although all in a relatively relaxed but alert range.

"Terrific! I think we're communicating with it," Lar said. "It's mimicking our transmission pattern, like it did with the ship."

Over the next 15 min, Sam tried several different combinations of intensity and frequency, and each time, the response was similar. The EEG showed a parroting effect, but always in the relaxed to slightly activated range.

"All right, let's reverse things and let it take the lead," Lars said. He went over to one of the exposed tangles of crystalline fibers near the perimeter and broke off a small piece with his hammer. He then returned to the EEG machine.

****

It became aware of some damage to its periphery.

"Danger!" it thought. But as it began to activate itself, it realized that the danger had stopped, and that it in turn needn't escalate its response. It quickly relaxed.

****

Glancing at the EEG, Lars commented: "Sam, it looks like it began to show a hyperalert response by abruptly increasing its frequency and wave amplitudes, but then it settled down and is now showing a more relaxed pattern."

"And I experienced a sort of panicky feeling for a moment," she responded, "but that's now gone away."

"I think you reacted to the higher amplitudes. Try to match the pattern with your generator."

Samantha broadcast a high frequency pulse with high power in the radio spectrum, then she started to lower it slowly.

"It worked," Lars said. "The network mimicked your pattern with an abrupt high frequency/high amplitude wave, followed by a gradual lowering of both"

They repeated this sequence twice more, with a similar response from the network. Then they turned off the generator.

****

"What happened?" it thought. "Has it gone? Am I alone?"

It became concerned and started searching for the visitors."

****

"Sam, it didn't mimic us this time. It's showing some high frequency waves with a somewhat elevated amplitude."

"I'm getting an empty feeling, Lars. I think it's missing us."

"OK, go ahead and generate some mellow waves again."

She turned the generator back on to broadcast in the low frequency/low amplitude range.

"It's showing a more relaxed pattern again," Lars said, "around 9 Hz. And low amplitude"

"And I'm feeling more relaxed myself," Sam said. "Or rather, the part of me empathizing with the network is feeling more relaxed."

"I have an idea," Lars said. "Let's try pairing a pattern of five high/low frequency waves at standard intensity, then pause, then crank up the amplitude for a single pulse, then stop. Maybe it will associate the five frequency oscillating pattern with a danger warning."

"I get it. We're trying to teach it a frequency alphabet using our baseline intensity. It's worth a shot."

Sam activated the generator.

****

Something was coming from the visitors. A pattern. Then a frightening impulse. As it started to react, the impulse stopped.

"No danger now," it thought.

It reflected on what happened. Five high then low frequency waves, then a terrible activating wave.

"Were visitors trying to warn me of wave danger?" it considered.

It reflected on the possibility that the visitors were also aware, like it was. How to respond? Maybe it would try to communicate the same thing back.

****

"Something is happening, Sam. I'm seeing a saw tooth signal of five high and low frequency waves at its normal baseline EEG intensity, then a really high amplitude wave. But now the EEG shows the network's standard relaxed pattern."

"And I'm not feeling any fear or anxiety right now," Sam said. "I think it's conjuring up a reactive pattern as a signal, without really feeling an arousal reaction itself."

"Yes, I agree. It seems to be telling us that it knows the five-cycle high/low frequency baseline pattern that we sent it means danger is coming. But it doesn't feel threatened right now. It's telling us that it understands by mimicking our stimulation. I think we're communicating in a kind of frequency/amplitude Morse Code. Chuck would be proud of us!"

Samantha laughed.

"Now let's try a seven-wave oscillation pattern, followed by a low power wave."

Sam performed this action. The network responded similarly.

"The second letter in our network code alphabet," she said, laughing. "Five oscillations preface a high stress wave, and seven indicate that a low intensity wave will follow."

"Yes, we have two letters, now," Lars said.

He glanced at his computer watch.

"Whoa, I think we should get back to our rover. Our hour will be up soon. We can call the ship, tell them what happened, and await further orders."

"Sounds like a plan," she said, beginning to deactivate the generator. "And you know what? I'm feeling happiness coming from our friend."
****

It rejoiced. It was communicating with the visitors. Five high/low frequency waves meant danger was coming. Seven high/low frequency waves meant a calm period would follow. The visitors were aware and able to think, like it was. It was not alone. Praise the Maker for sending it companions.
****

"And that's where we stand with the network," Lars said, speaking into the *Rover 2* radio link to the *Hermes*. The two of them had just returned and had broken out their rations. Lars took another bite from his sandwich.

"Very good," Kilborn said from the other end. "You made good contact with this network of yours, and that should make the job easier for the next crew."

Lars glanced over at Samantha, who suddenly looked glum.

"So. . .what are our orders."

"Get back here right away. Chuck estimates being finished calibrating the computers and checking out the electronics by midnight tonight. We have a launch time scheduled for 0200 tomorrow morning. Anthony and I plan to get a little shut-eye after dinner. It wouldn't hurt for the two of you to do the same."

"Captain," Samantha interrupted, moving closer to the speaker. "Do you think we can delay the launch? We're making real progress with the network, and I think. . ."

"Sorry, Samantha. I gave you a chance to test some of your ideas, but now I have to consider the safety of this mission," Kilborn interrupted. "So long as we have a window to launch, we should take advantage of it. Who knows what this alien will do next. It has the ability to cripple us at any time. My orders were to find out what caused the mysterious signal and to return with that information. We've already lost two crewmembers, and I don't plan to lose any others. So the two of you. . .get back as soon as possible."

"OK, Captain, will do," Lars said. He glanced at Samantha as he signed off. The two of them began to gather their equipment.
****

Samantha and Lars returned in the late afternoon. They drove *Rover 2* up the ramp directly into the cargo hold, where they parked it next to *Rover 1*. Someone had earlier moved it there for the launch. They secured their rover in preparation for later disassembly during the long trip home. They then sealed off the cargo hold from the outside, went up the outside elevator to the airlock, entered the ship, and headed for the space suit stowage room. They took off their suits and found their crewmates in the mid-deck, where they were preparing dinner.

"Well, you certainly had a nice chat with the alien," Kilborn said.

"Yes, Sir, we did," replied Samantha. "We started to develop a primitive type of network Morse Code communication. I think it that it understood our intentions. This is all very exciting."

"Wonderful," Chuck said. "This bodes well for future missions to Mercury, so long as the network perceives these missions as peaceful."

"I hope we can leave it with a memory that we're its friend," Lars said.

"Captain, we're ready whenever you are for the countdown," Anthony announced as he came down from the fore-deck. "I have finished the final checks. All launch systems look good."

"Terrific. Well boys and girls, let's have a hearty dinner, take a nap, and plan to launch at 0200 h tomorrow morning."

They ate in silence, then withdrew to their sleep pods. Samantha tossed and turned, thinking about the network and how they were just starting to scratch the surface of its consciousness. She was saddened about leaving so soon.

This is a momentous event, she thought. *Astrobiologists throughout the Solar System will have to change their views about carbon-based organisms being the only viable form of life in our universe. The network shows that silicon-based life and perhaps life based on other novel chemistries are possible.*

Her mind began to wander about what such other life forms might look like. Before long, she drifted off to sleep.

Mercury, Day 75

The ship's alarm awakened the crew at 0100 h. After dressing and grabbing a quick snack, they headed to the fore-deck. Samantha, Lars, Tilda, and Chuck went directly to their acceleration couches. Anthony and the Captain entered some final launch instructions at the central control console, then proceeded to their own couches. They swiveled their seats to the correct orientation for launch and telescoped their personal computers out so that the screens were in front of them.

After entering a series of commands, Anthony announced: "Launch minus 20 min!"

As the various systems were activated in proper sequence by the central computer, the Captain scanned his crew.

"Well, we'll soon be off. I want to say that it's been a pleasure working with all of you. We had some tough challenges, but we managed to survive them. We certainly accomplished our mission goals."

"Thank you, Sir," Chuck said from his couch. "It's been a pleasure."

The others concurred.

As the minutes ticked down, the fission reactor began to power up the engines by thermally heating the stored hydrogen that would provide the thrust for their launch. With this activity, some radiation leaked out of the system to the outside. It did not go unnoticed.

****

It perceived a change. Suddenly, new waves were being transmitted from the distant intruder that had brought the visitors.

"What's happening?" it thought. "Is it danger? No, the signal is remaining distant."

Then, another thought.

"Are they leaving? Will I be alone again?"

Alarmed, the network started to react. This was not a danger reaction. It was one of panic. True, the Maker was still above. But now that it knew there were other conscious entities in existence, it again began to experience loneliness.

****

"Launch minus 2 min!" the central computer announced over the intercom system.

"Captain, something is happening," Tilda said, checking her computer screen. "The external radiation level is increasing again."

"Is it from the fission reactor?" Kilborn asked.

Tilda responded: "No, Sir. The radiation is coming from the network."

"Launch minus 1½ min!" the computer announced.

"How can that be?" Kilborn wondered.

"Maybe it doesn't want us to leave," Lars added.

"Oh, damn, don't tell me that thing is interfering again!"

Maybe we'll be staying after all, Samantha thought.

"Captain, some of the systems are starting to shut down," Chuck said.

"Launch minus 60 sec. . .59. . .58. . .57. . .56. . .", the computer continued.

"Which systems?" asked Kilborn.

"My screen indicates that the central computer is losing its integrity. Life support is holding, but the reactor and hydrogen flow performance monitors are in the yellow zone."

"Launch minus 46. . .46. . .52. . .49. . ."

"Chuck, can we launch? The computer is giving faulty readings."

"I don't think so, Captain. The reactor is starting to shut down."

"I don't believe it," said the Captain. "Anthony, abort the launch!"

"Aye, aye, Sir," he said, as he initiated the emergency shut down procedures.

****

The radiation began to decrease from the intruder. It had learned that this meant that it too should decrease its own radiation. Deep down in the ground, ionic fluids under the network were diverted into channels that were cooler, and this slowed down their motion. As a byproduct, the EM waves that radiated outwardly began to decrease in intensity. The network relaxed.

****

As the countdown was being aborted, Tilda announced: "Captain, the radiation from the network is decreasing. We're heading toward our pre-countdown baseline."

"Well, everyone," the Captain said, "I guess we're going to be here a while longer."

After the crewmembers completed their individual shutdown procedures, they each got up from their acceleration couches and proceeded to the mid-deck. Anthony stayed behind to complete the post-abort protocol and survey the ship's status. The rest of them gathered around the dining table for some tea and a discussion.

"It would appear that the alien doesn't want us to leave. Any ideas why?" Kilborn asked.

"Captain," Samantha said, "for a moment during the countdown, I felt a strange anxiety, then a fear of being alone. It was a weak feeling, and fleeting, but definite."

"Oh great, it wants some buddies to play with," Kilborn said.

"I think it's more than that." Lars said. "It's probably curious about us. Who are we? What are we doing here? Plus, who knows how long it's been conscious? Maybe millions or even billions of years, waking up in the daytime to nothing but itself and the Sun, going dormant at nighttime, all alone. Now that it's sensed other entities on Mercury, it likely realizes what it's been missing. If you were the network, wouldn't you want us to stay?"

"I guess so," Kilborn said grudgingly. "It looks like we'll just have to hang around until this entity goes dormant again. Then, with no further interference, we can take off. Tilda, what are your heat and radiation projections?"

"Captain, with the Sun sinking ever lower toward the horizon, and the external temperature dropping as a result, time is on our side. The average daily radiation levels coming from the network have been decreasing. So long as we don't excite it, I think we'll be fine. The heat and radiation shielding system is working nominally."

"And the electronics and computer programs are also functioning well," Chuck added. "I just need to tweak a couple of programs further, but everything should be perfect in a few days."

"All right," Kilborn said. "I'll notify the Space Council of our status."

"And please have them tell the Space Services to lay off any plans they have to rescue us," Samantha said. "They could just cause more trouble."

Kilborn nodded. He then turned to Anthony, who had just entered the mid-deck from the elevator.

"How are our consumables holding out? Can we remain on the surface awhile until the alien goes to sleep?"

"Yes, Sir. With some rationing, we should have enough emergency food, water, and oxygen to cover our stay until the beginning of local nighttime, and still get us home. When you decide it's time for us to leave, I can get everything ready in 24 h, assuming the network leaves us alone."

"We should be able to launch before the Sun sets in 13 days—or whenever the network goes dormant," Kilborn said.

He turned to look at Lars and Samantha, who were sitting next to each other.

"I guess the two of you will be able to learn more about our alien friend. You'll have some time to have a really nice chat with it."

"Aye, aye, Captain," Lars said.

Samantha beamed. Her mind started working on ways to further study the network.

Mercury, Day 76

Admiral Harvey was in his office reviewing the report of the previous day's failure of the *Hermes* to launch. The holophone alarm went off. It was his secretary, announcing that General Suzuki was outside and wanted to see him.

"Send him in," Harvey responded.

Now what, he thought. *Another crisis?*

"Good afternoon, Admiral," Suzuki announced as he came through the door.

"Hello, General. Have a seat. What can I do for you?"

"You are aware that the alien prevented the *Hermes* from launching yesterday?"

"Yes, I am. I have the official report in front of me."

"Do you have a plan to recommend to the Space Council?"

"Not yet. I was going to call an emergency meeting for later today."

"Before you do that, I want to discuss something with you. The Army has been interested in employing a new holotransmission system that uses focused neutrinos as a carrier to send the message. We think that it will be useful in transmitting messages through surface rocks and hills in land operations where there are no satellite relays. Major Chang has been testing this system outside the *Hermes*. She has been sending messages from behind barriers and relaying them to us via the ship's transmission system. It looks promising."

"Wonderful news, but what has this to do with launching the ship?"

"Well, the neutrino beam is produced by a portable nuclear reactor that is part of the transmission system. This reactor could be made unstable, essentially becoming a small bomb. If it could be carried close to the alien and exploded, it may solve our launch problem."

"Interesting notion. But how would this happen?"

"We could order Major Chang to take the transmitter to the plateau, allegedly to test the system from a great distance away from the ship. Once there, she could take it close to the network, set the reactor to explode on a timer system, and poof! Our troubles would be over."

"Very clever idea. Could Major Chang carry this out?"

"Yes, we just need to brief her on a secure channel. Being the comm officer on the ship, no one else would know of our transmission to her."

Harvey leaned back and smiled.

"According to the report, the crew thinks the alien will become dormant as the Sun sets, and they can launch then. But this thing is unpredictable, and it is a danger to any of our ships flying near Mercury."

"My plan would let us deal with this thing once and for all," Suzuki said.

Harvey again consulted his report.

"The crew is planning another visit to the plateau in 2 days. This would give us time to run the plan by Santini and iron out the details. I don't see any reason to involve the civilians on the Space Council. This would be a purely military operation."

"I agree," Suzuki nodded.

Harvey reached for the holophone to call General Santini.

Mercury, Day 78

On the 78th day of their mission, the temperature had dropped to 300° as the Sun continued to move toward the western horizon. Tilda reported that the signal from the network was barely detectable.

Just after breakfast, the Captain announced that Tilda would be joining Chuck, Samantha, and Lars on their trip to the plateau in order to conduct a transmission experiment under orders from the Army. Tilda briefly described her experiences near the ship with the neutrino transmitter and the decision to test it from a further distance away. Samantha noted that Tilda seemed preoccupied as she conducted the briefing, as if there was something else on her mind.

Their trip to the plateau was uneventful. After arriving, they parked *Rover 2* and headed up to the top, taking not only the EM generator and the modified EEG machine, but also a special portable drill to look for minerals under the surface that could be used for future mining activities. Tilda also wheeled up her transmitter, a feat she was able to handle in the low gravity of Mercury.

At the top, Lars and Samantha went over to the network to hook up their machines, while Chuck activated the X-ray machine that still remained on the plateau and began imaging and drilling under the surface. Tilda set up her transmitter closer to the network.

"The EEG shows slower waves," Lars said over the intercom after turning on the machine.

"Yes, that makes sense," Samantha responded. "The Sun is lower in the sky, providing less heat and fewer photons to activate the network. I suspect that within the next week it will become pretty dormant, and the silicon will begin to crystalize, allowing more material to be added to the rods and more connecting fibers to form. Curious how the network grows when its metabolic processes are slowest."

As she said this, she sensed a presence in her mind, this time curiosity and comfort with their being there. There was a general sense of contentment. She tried to relax, projecting a sense of comfort as well. To her surprise, she felt a response of joy. She and the network were interacting on an emotional level! She kept quiet about this and continued to set up the generator for their subsequent experiments.

"Whoa!" Lars exclaimed, examining the EEG readings. "The waves just increased a bit in both frequency and amplitude. Nothing for us to worry about, but the network seems to have become more alert. What do you think is happening?"

"I think it's happy to see us," she responded.

****

It felt a new emotion, one of pleasant activation. One of the visitors was interacting with it. It had thought that only the Maker could do this. But the Maker could only cause it to form more of itself. This was different. The visitor seemed to be communicating with it directly, like one entity to another. This was new!

It became joyful.

****

"I found something interesting," Chuck announced over the intercom. He was standing about halfway between the network and the rim. "The X-ray is showing some kind of space under here. It looks like a tunnel. Maybe it's an old volcanic vent."

He activated the portable drill as Samantha and Lars went over to him. Tilda remained behind with her transmitter.

After 10 min of drilling, the drill tip broke through to an open space.

"The temperature down there is a cool 160 °C! The space seems to be empty—there's no lava. Who knows how far it extends. We need to take a closer look at this. Maybe there's something interesting to discover."

Looking at the X-ray screen, Samantha asked: "Do you think the network is involved?"

"No, I don't show any signs of anything extending into the space. No fibers or solid masses. It's just a hollow tube. But it's leading down to something further below the surface."

He activated the comm line to the *Hermes* and made his report.

"Interesting," Kilborn said. "Let's leave the exploration of this space for another day. Complete your talking session with the alien, then return to the ship."

Lars and Samantha went back to their stations. They generated some additional wave patterns and in the process expanded their network code alphabet. After 25 more minutes, they decided to call it a day.

****

While Lars and Chuck were shutting off their equipment, Samantha walked over to Tilda.

"We need to leave soon. How are things going?" she asked.

"Fine," Tilda responded tersely. "I will be finished in a few minutes. Go back to your equipment, and I will follow you shortly."

Samantha glanced over at the transmitter and noted an attached clock showing progressively decreasing digital numbers.

"What's that?" she asked.

"Nothing," Tilda responded. "Leave me alone to do my work."

"It looks like a timer. What happens when the numbers reach zero?"

"I said leave me alone, you snoopy bitch!"

With that, Tilda activated a switch, stood up, reached into her shoulder bag, and pulled out a laser pistol.

"Back off!" she said.

Samantha backed away, noting at the same time that the altercation had attracted the attention of Lars and Chuck.

Tilda glared at her through her helmet face plate.

"You always defend this alien thing, you and your doctor lover boy. Well, the Army sees it as a big threat. They ordered me to blow it up, which I am doing. My transmitter is now a bomb. In less than 5 min, there won't be much left of it."

She looked at the others.

"We all had best get out of here unless we want to blow up with the network."

"This is stupid," Samantha said. "The network is becoming dormant. You yourself could barely pick up a signal from it back at the *Hermes*. In the next few days, it won't be much of a threat to us. We can take off safely."

"Shut up!" Tilda retorted. "I am under direct orders from my superiors. You can't do anything to change my mind."

Samantha started to move forward toward Tilda, hoping to disarm her and deactivate the bomb. Tilda fired her weapon. The beam hit the ground just in front of Samantha.

"Keep away, I said!"

Samantha abruptly stopped and moved back again.

****

In its drowsy state, it felt new emotions, which caused it to activate.

"Something is wrong. There is danger."

It perceived two presences in conflict. One visitor, the one that had earlier caused it joy, was reacting with fear. The other one was reacting with anger and was threatening the first in some way. Near the second visitor was a radiation leak that was like the ones it had perceived before when it was attacked.

"Must protect myself. Must protect my friend."

It began to focus what little energy it had left.

****

Samantha was trying to decide what to do, when suddenly a flash erupted from the network. She was knocked down by the concussion. As she stood up, she saw that Tilda also had been thrown to the ground, her laser pistol lying several feet away. Tilda started to get up, eyeing the pistol. Samantha bolted up and propelled herself to the weapon first.

"I won't let you do this!" she said.

She turned and fired several times at the transmitter, destroying the timer and much of its mechanism. She then turned the pistol on the groggy Tilda.

"Samantha, don't!" Lars yelled from behind.

For a moment, she felt a protective rage at Tilda, much like a lioness feels when her cub is being attacked by another animal. Then, just as suddenly, she repressed the urge to fire, realizing that the threat to the network had disappeared. She lowered the gun.

"Your bomb is destroyed," she said. "Let's go home."

Lars and Chuck arrived on the scene. While Chuck helped Tilda up, Lars took the weapon from Samantha. He then performed a quick vital sign check on the two women with his medical scanner.

"You both seem OK," he said, turning to go back to the rover.

As they made their way across the plateau, Samantha glanced back at the network. She perceived a sense of peace from it as it returned to its slumbering state.

****

During the trip back, Chuck reported the incident to the *Hermes*. Since Tilda was acting on military orders from the Army, and since her activities were in support of her mission, the Captain didn't feel that she was a threat to the crew. He commented that the incident would no doubt lead to a major altercation within the Space Council, but this was not their concern. He also commented that Samantha would have some explaining to do, but as a civilian, she was not bound by military orders.

"Just come back to the ship and we'll go from there," he said.

****

By late afternoon, Lars had medically cleared Tilda and Samantha to perform light duties, and they all gathered for dinner. Surprisingly, Tilda apologized to them for her actions, saying that she was just following orders. They accepted her apology, although there were a few snide comments about the military from the civilians.

Later that evening, Samantha heard a knock at her sleep pod door. It was Tilda. She invited her in, and they both sat down.

"I'm really sorry about what I said to you up at the plateau," Tilda said. "At one time I had a crush on Lars, and I guess some of my old feelings just spilled out. But I realize that you and he are an item, and I can live with this. In fact, you seem to be well suited for each other."

Samantha didn't know what to say and remained silent.

Tilda continued.

"But there's something else. As a communications person, I really respect what the two of you are doing with the network. I have been quietly supporting your actions, despite some of the opposition from my fellow officers. When I got the orders from my superiors to bomb the network, it put me in a real dilemma. I really didn't want to see it destroyed. I guess some of my anger on the plateau was really anger at the actions I was forced to do. Between us girls, I'm glad that the network still exists, despite my still having a headache from its energy blast."

Samantha smiled, stood up, and gave her a hug.

Mercury, Day 81

Two days later, the temperature had dropped to 250°. Another excursion to the plateau was scheduled for that morning. As Samantha fired up *Rover 2*, Chuck and Lars went over to one of the rear supplementary supply holds to retrieve two special portable power drills that were equipped with laser projectors that could liquefy the regolith. They had spent the previous day testing the drills in preparation for their activities on the plateau. Everything had worked perfectly. They carried both drills back and stored them in the rover.

The three of them left on their excursion. After they arrived, they parked the rover and carried their equipment up to the plateau. Samantha took the EM generator and the EEG machine over to the network. Lars remained behind to help Chuck dig down into the ground over the location of the tunnel that Chuck had discovered earlier.

With the help of the laser drills, the two men created a hole large enough for them to fit through. They continued to dig down until they broke through the top of the tunnel. They attached a portable elevator ladder to the side of the surface hole using pounded spikes, turned on their helmet lights, and lowered themselves into the tunnel. Their helmet lights revealed that the pathway curved downward at a sharp angle.

"I wonder how far we'll go until we reach the end?" Lars asked.

"If I'm correct in assuming that this is an old lava tube, it's hard to say," Chuck responded. "It must be blocked off, or else we would be frying in molten lava."

After walking for some time, they entered a large cave some 40 m across and nearly that amount in height. There was no other visible opening signaling a method of egress, although across the way Chuck noted a clump of rocks that may have been an old landslide.

"I'll bet there's lava behind those rocks," he said.

He walked over to the area and touched the rocks with his digital environmental scanner.

"Yup, they're warmer than the adjacent wall rocks. The lava that's likely powering the network is probably located behind this barrier."

"And look over to the right," Lars said, pointing. "I think there's some gaseous substance rising up from those vents."

Walking back to the center of the cave, Chuck continued to monitor his scanner.

"The pleasant surprise is that the temperature down here is a balmy 30 °C," Chuck said. "The rocks surrounding us must be insulating this space from the tremendous radiation and heat above at the surface during the local Mercurial day and the freezing cold during the night time. Whatever heat source might lie behind the rockslide seems to be transmitting enough heat through the rocks to give us a stable livable temperature."

He paused, a shocked look on his face.

"Hold on! I'm getting other readings on my scanner. You're right about the gasses. I'm finding carbon dioxide, hydrogen sulfide, nitrogen, argon, methane...even some water vapor. There's an atmosphere in here. Not breathable, but enough to produce a substantial pressure."

"What can be producing these gases?" Lars asked.

"My scanner is telling me that if we were on Earth, these gases would be characteristic of those that are associated with volcanoes. The vents must be leading to a hot magma source down below."

"That would make sense," Lars said.

The two of them walked over to the gaseous vents.

"Yes, the atmospheric pressure is increasing over here."

"You know," Chuck continued, "this cave would make for a nice colony site, especially with the comfortable temperature and pressure. You would need to stay in an oxygenated enclosure, but you could probably explore the cave in shirt sleeves and just an oxygen mask."

"Sounds like a good recommendation for a future mission," Lars said, checking his computer watch. "Well, we've been here long enough. Let's take a few holopictures and head back up and see what Sam's been up to."

****

Samantha had been busy. She first had set up the EM generator, where she had entered a pre-programmed protocol for delivering stimulation to the network in keeping with elements of the network code system that she and Lars had worked out. Finishing this, she walked over to the network to hook up the EEG machine. A thought came into her mind. It was weak but distinct. It made her sleepy. A glance at the EEG showed that the network was displaying its characteristic, relaxed 10 Hz waves, but that more synchronized, relatively high amplitude 2 Hz waves would occasionally intervene. Samantha recalled Lars telling her that these were the kind of waves found in sleeping humans. The network was beginning to enter its dormant period.

Then Samantha had an idea. She sat down on a large rock nearby and began to think hard about a number of positive things. She tried to project peaceful thoughts, thoughts about friendship, about loyalty, about caring. She sensed calmness coming from the network. A notion came into her mind that she was needed and desired. She responded by thinking about leaving Mercury in a few days. The reaction was a feeling of sadness and longing. She imagined being gone, then coming back, and the response was a feeling of happiness.

Her reverie was interrupted by the sight of Chuck and Lars coming up from the hole they had dug. She stood up and began moving toward them. Instead of happiness at seeing them return, she felt increasing sadness as she walked away from the network. She stopped, turned back, and formed a thought that she would always be its friend. Joy entered her mind.

****

The Maker was leaving, the light was fading, and it was beginning to enter the quiet period. Just like billions of times before. Soon, new material would harden below, and it would grow in the process. It would note this the next time it awoke. All was as usual.

But something was not usual, for it had another companion in addition to the Maker. This companion could do something special that the Maker had never done—it could communicate directly with it in a kind of give and take. The companion indicated that it might have to leave physically but that part of it would always be there in memory. The notion of friendship entered its consciousness. It would always be thought of, even when it slept after the Maker left. This was a joyous notion. As it prepared to drift into slumber, it knew contentment.

****

The *Rover 2* returned to the *Hermes* in the late afternoon, so the formal briefing was deferred until dinner time. As they finished their meal, Lars and Chuck began to describe the cave they had discovered and presented their ideas on how it could be used as a base for a future mission. Samantha reported on her interactions with the network. She said that both her own subjective experience and the EEG data suggested that the network was becoming dormant. Tilda added that the ELF waves coming from the network had essentially disappeared.

"This is certainly good news," Kilborn said. "Maybe we can take off before the Sun sets, perhaps in a few days. Let's give the alien time to really fall asleep so that it doesn't bother us again during our launch countdown. I suggest that Chuck, Lars, and Samantha go back to the plateau the day after tomorrow. Tilda tells me that the outside temperature should be down to about 200° by then. At that temperature, it surely will be cold enough for the alien to be fully dormant."

He looked at Lars and Samantha.

"During your visit, you two can do your EEG number to verify that it's asleep. If so, we will still have some daylight left to secure everything and hopefully launch without interruption. Agreed?"

Everyone nodded.

Mercury, Day 83

On the morning of Day 83, Samantha, Lars, and Chuck left in *Rover 2*. They arrived at the plateau just before lunch. After a quick snack and trek up the path with their equipment, Samantha applied the EEG to the network. It showed synchronous, high amplitude 1 Hz waves, and Samantha felt no emotional input coming from the outside. She broadcast this information via the rover back to the *Hermes*. Tilda responded by reporting zero extraneous radiation from the plateau. So it appeared that the network had fully entered its dormant period. Chuck X-rayed the crystals and fibers located underground, and he thought that he detected some additional growth due to new crystal formation.

The three of them then went over to the hole that was dug earlier. They descended to the tunnel, then trekked along it until they reached the cave. Chuck and Lars began to walk around the perimeter to look for additional vents. While they searched, Samantha went over to the steaming area noted earlier. She ran some tests and verified that gases were being vented into the cave. She sat down on a rock and looked around. She marveled at the fact that this space was pleasantly warm and insulated from the outside. As Chuck had speculated, it would indeed make a good base for a future crew. It was well-situated to study the network, due both to its proximity to the entity and to its location underground, where the crew would be protected from solar heat and radiation. She also thought that they would be safe from the radiation generated by the network. The crew would have to sleep and work in a sealed enclosure, something like the rover tent, but the temperature would be fine, and when they made excursions out into the cave, they would only need an oxygen mask.

After half an hour, Chuck and Lars joined her. They reported finding a few more steaming vents, which likely contributed gases to the tenuous atmosphere. Then, the three of them went back to the surface. Chuck and Lars carried over a metal plug that they had engineered on the *Hermes* and brought with them in the rover. It had the same circumference as the hole, and they put it in place. This was to seal in the atmosphere so as to maintain the livable pressure in the cave.

As they walked over to the rim of the plateau, Samantha took another look at the network. The crystalline rods cast long shadows from the Sun, which was located low near the western horizon and would be setting in 5 more days. The three of them walked down to the rover and drove back to the *Hermes*.

Mercury, Day 84

"Absolutely not!" Kilborn bellowed. "We're done exploring. Now that the alien—the network—is out of commission, we're launching off this rock and going home. We plan to leave in 2 days. That will give us enough time to retrieve the perimeter radiation detectors and other stuff we've left in the neighborhood, plus secure our supplies and equipment and stow the rovers back on board the ship. The next crew visiting Mercury can continue our mission, now that we know what we're dealing with."

"But Captain," Samantha said over breakfast, "we have enough daylight left for a quick 4-day round trip excursion up north to investigate some of the icy craters. It's about 2400 km to the nearest one. If we drive non-stop, we can make it there in 2 days, explore a little bit and collect samples, then come back before nightfall. Who knows what we'll find buried in the ice? Perhaps multicellular organisms, like those discovered in the subsurface of Europa. Or perhaps some new kind of life form."

"That's cutting things too close," Kilborn responded. "I don't want to postpone the launch date, and I don't want any of my people gallivanting around the planet that close to sunset. If you're delayed for some reason, you'd have to return in the dark. Plus, I would just as soon launch when there's a little bit of daylight."

He poured himself another cup of coffee.

"No, we'll leave further exploring for the next crew."

Samantha was disappointed. Now that the network was dormant, it would be fascinating to look for life in some frozen water. She understood the Captain's position, although she personally thought a trip would be worth the risk. So what if they arrived back after sunset—the temperature would still be high, and the rovers had lights. Nevertheless, the Captain was in charge. But still, it got her thinking. . .

****

Just before lunch, Samantha sought out Lars. They met in his sleep pod, allegedly to review lab tests stored in Lars' computer. Some of the other crewmembers suspected a romantic intention, but that was far from the truth.

"So Sam, what's on your mind?" Lars said as he closed the door and sat on his bed. He directed her toward the chair in front of the computer table.

"I'm so disappointed that the Captain won't let us explore an ice crater before we leave. This is a nice opportunity to take another look for carbon-based microorganisms that need water."

"Yes, it's too bad. But he wants to launch under optimal conditions. And the craters could be explored by future expeditions to Mercury. There likely will be several such missions, especially in light of the existence of the network. You could apply to be included in the next mission and are sure to be chosen, given your experiences during this expedition."

"I don't know…there's no guarantee that I would be selected. Plus, I'm worried that the Space Services might do something to hurt the network. Space ships need to go by Mercury to get gravity assists, from the Sun and there's always the danger that one of them could be perceived as hostile by the network and get zapped. The Space Navy again may decide to do something to prevent this possibility."

"I don't believe that the Space Council will allow this to occur. The network is something special. You and I have started to communicate with it, and we know that it has awareness and possibly some intelligence, if we can talk about such a thing in a crystal."

He smiled at this thought.

"Yes, that's my point," Samantha said. "You and I have made a great start, and we're in the best position to continue exploring the utility of our network code system in expanding this communication."

Lars thought about what she was saying. Then he looked at her.

"Sam, there's something else you want to tell me, isn't there?"

She looked down at her hand resting on her lap, then up at him.

"The last time we were on the plateau, and you and Chuck were in the cave, I went over to the network to do some testing. I got this feeling from it, and so I just sat down by the EEG machine. Lars, the network was talking to me. Or at least projecting its feelings. And I could respond to it. It's lonely. It recognizes me as another entity, as something it can relate to, something to keep it company."

She paused.

"It's been alone for, I don't know, maybe billions of years. And here we are. And it knows what…who…it has been missing all these years. I projected a thought to it that I was its friend, that I would always be with it, at least in thought. It reacted with joy. I think it went to sleep happy. I believe that it expects me to be here when it next awakens."

"But will it remember you at the next awakening?"

"I think so. During the Mercury night, when things cool down, part of its newly formed crystal lattice could store the preceding daytime experiences, in a manner similar to the way experiences are encoded and remembered in the hard wiring of our brains. When things heat up during the next Mercury day, and current begins to flow again, these memory traces could be re-experienced in its consciousness. I think it would remember me and again react with joy."

"Jeeze, Sam, you sound like its lover," he laughed.

"No, you're my lover, Lars. The network and I have something else going for us. It's like what I felt with my father when he was in the nursing facility. He seemed happy to see me, but we weren't able to communicate except in non-verbal ways. It's hard to explain. There was an emotional connection."

"I get it, Sam," he responded. "Mothers and babies have this kind of attachment. They can't talk to each other, but they nevertheless communicate."

"Right."

She paused and looked down ruefully, collecting her thoughts. Then she continued.

"Lars, I lost contact with my father when he developed Alzheimer's. Part of him must have felt terribly isolated and lonely, especially as he descended deeper and deeper into his disease. I can't lose contact with the network. It has awareness, and I feel like it needs me."

"I understand," he said.

He stood up and gave her a hug, feeling her warm cheek against his. After a time, they separated.

"So, what's on your mind?" he asked.

She looked directly into his eyes.

"I have a plan," she responded.

Mercury, Day 86

The ship's alarm woke the crew up early on the morning of Day 86. The previous day had been busy, with everyone pitching in to retrieve the radiation detectors and other equipment that had been placed outside the ship. After breakfast, the Captain and Anthony planned to test the systems related to the launch, scheduled for later in the day. Tilda was to check out the communications system, then leave her console for a few hours to help Chuck verify the integrity of the central computer and inspect the fission reactor. Lars and Samantha had volunteered to go outside, prepare the two rovers for storage, and move them into the cargo hold. Then they were to fold up and store the rover tent.

At lunch time, the Captain, Chuck Tilda, and Anthony gathered to make their final assignments.

"Where are Lars and Samantha," Kilborn asked?

"Oh," Tilda said, "they left word for me to tell you that they would be eating on *Rover 1* since they didn't want to have to come back in to the ship, doff their space suits, eat lunch, then don their suits to go out again. They said they would call us to report in after lunch."

"Maybe they just wanted one last time to be alone," Chuck said, smiling. "It can be pretty intimate in the *Rover 1* cabin. Just turn off the comm, and no one knows what's going on inside."

Everyone laughed.

"OK, let's leave them alone," Kilborn said. "Now, the rest of you, let's go over the assignments for the rest of the day."

****

By 1400 h, there still was no word from Lars and Samantha.

"Contact them for a progress report," Kilborn said. The crewmembers were all in the fore-deck reviewing their respective launch protocols.

"I have a visual outside," Tilda responded from her console. "No sign of the rovers or the tent, so they must have already stowed everything."

"That's curious," Chuck said. "I don't remember seeing the keyboard lights flash on to indicate the opening of the cargo hold door and the descent of the ramp. Let me check something."

After consulting his computer terminal, he reported: "Captain, the rovers are not in storage. There's no sign of them, except that the location signal shows both moving west about 40 km away."

"What the hell are the two of them up to?" Kilborn said. "Tilda, get them on the comm."

"I'm trying, Sir, but there's no response. Let me check for messages."

"Damn! I bet they're going to the plateau. I want to launch in a couple of hours. If they delay me, I'll lock them up in the aft-deck storage room during the trip back."

"Captain, there's something you ought to see. It's a message sent out last night by Samantha to her mother. It contains some interesting information."

"All right, project it on the main holoscreen."

The four of them moved forward to read what it said.

****

To: k.evans@stationdelta.org (RESTRICTED USE)
From: samantha.evans@Mercuryexplore.assc
Subject: Goodbye
Date: Mercuryexplore, Day 85

Hi Mom,

This will be my last message for a while. As I wrote earlier, we have found an underground cave near the network that has a livable temperature. The space also has some atmospheric pressure due to the expulsion of volcanic gases from below via several vents. It's a good environment for people to live in, provided they have some oxygen to breathe.

After careful thought, Lars and I have made a decision. We plan to take the two rovers and remain on Mercury, using the cave as our base. It should be quite comfortable. We will live in the large tent that originally was used to house the two rovers. It can be pressurized and sealed to retain oxygen. With proper rationing, we'll have enough food and water to sustain ourselves for nearly a year. We also will have a small nuclear power generator, a supply of oxygen canisters, and an air supply generator that will let us produce more oxygen from the carbon dioxide that is located in the cave. We plan to retrieve these supplies and equipment from the *Hermes* supplementary supply holds during our pre-launch inspection tomorrow, and then put them in the two rovers. After folding and storing the tent on *Rover 2*, we'll leave for the plateau.

When we finish setting everything up in the cave, our plan will be to make a trip in one of the rovers to an icy crater located up north. Although it will be nighttime, it should be warm enough outside, since the daytime heat will not have had enough time to radiate completely away from the surface. We'll take along some extra charged batteries to power the rover, and we'll be careful and take our time since we'll be traveling in the dark. But the rover has ample lighting for night driving. Depending on what we find, we'll return with samples from the area, along with some ice to restock our water supply. We plan to conduct scientific studies after we return, as well as write our reports and explore the cave even further.

When the Sun rises again in 3 months and the network wakes up, we'll continue our attempts to communicate with it. I've been able to sense its emotional reactions to things, and I believe it can do the same with me. I don't understand how this occurs, but it does. It has something to do with our brain waves resonating with each other. Anyway, I feel like we're friends, and I want to continue to explore this relationship in the future. I know this sounds strange, but that's the way I feel, and I believe that when the network wakes up, it will be pleased to sense that Lars and I are around and that it's not alone.

So, Mom, I'll say goodbye for now. I'm sorry to be apart from you for a longer period of time than we had anticipated, but it sounds like you have a nice group of friends, and I'm sure you'll continue to develop even more relationships. I expect that Luna City will be sending another expedition here in the near future, and perhaps Lars and I will have finished our work by then and can hitchhike a ride back with them. I expect to see you in a year or so. And who knows, at that time, you may be getting a new son-in-law. Stay well!

Love,

Sam

PS—I plan to name our cave home after Dad: *Dylan Evans Base*. I don't know if the name will stick when future crews start coming to Mercury, but I hope it does.

****

"Well, I guess we know what their plans are," Kilborn said.

"Captain, we're getting a message from *Rover 1*," Tilda said.

"You might as well put it on the intercom," he responded glumly.

"Hello, *Hermes*. This is Samantha, broadcasting from *Rover 1*. I have an open channel to Lars, who's driving *Rover 2*. Can you read me?"

"Yes, we can," Tilda said. "We're all here."

"Good. I sent a message to my mother last night which explains what Lars and I are up to."

"Yes, we just read it!" Kilborn interrupted. "But I'm not very happy about it. I want the two of you to turn around and come back immediately. I order it!"

"Sorry, Captain," Samantha said, "but we plan to stay here until relieved by a future crew. By then, we hope to be talking with the network and to have some notion of what the ice craters have in store for us. We expect to conduct some interesting science over the next night and day on Mercury."

"But you can't do that! It's not in our mission statement to set up a colony. Lars, talk some sense into her."

"I agree with her plan," Lars answered. "Sam is fully qualified to conduct the science she plans to do. She has taken the lead in communicating with the network, and I will be an enthusiastic assistant. Plus, with the rovers available to us, we expect to have some interesting experiences exploring the surface."

"You are disobeying orders. You must come back!" Kilborn persisted.

"With all due respect, Sir, we must refuse," Lars responded. "I guess we'll pay the price when we're relieved of duty by a rescue crew. But perhaps people will be lenient when we share our findings with the scientists in the Space Council."

There was silence from the *Hermes*. Then the Captain spoke again.

"I guess that's all that can be said for now about this incident. You have the two rovers, and we can't very well come and get you. Just take care of yourselves."

"We have everything we need to survive," Samantha said. "I hope you have a successful launch and an uneventful trip back to Luna City. When're you leaving?"

"At 1700 h this afternoon," Anthony added. "I have to program the launch parameters to account for the lesser mass due to the absence of the rovers, the two of you, and whatever supplies and equipment you took from the rear storage compartments."

"Yes, Anthony," Lars said. "We did make a raid on things this morning. And you probably know that we took the rover tent with us. It should be a comfortable home for us in the cave."

"Captain..." Samantha said, "...actually all of you...it was a pleasure serving with you on this mission. It has turned out to be quite an experience, and I'm glad to have been included, even though some people thought that having an astrobiologist in the crew was superfluous. I hope you have a safe journey to Luna City. God speed."

"The same goes for me," Lars added. "I enjoyed knowing you, and I hope you have a safe mission back. Since all of you have had some basic medical and cardiac resuscitation training, you should be able to deal with any aches and pains that may occur during your return, without a ship's doctor."

"I wish you a safe stay on Mercury until you're picked up," Tilda said. "I'm sure you'll have interesting experiences at the ice crater as well. Good luck."

"Same goes for me," added Anthony.

"Me too," said Chuck. "And I see that the two of you have finally learned how to drive the rovers. Watch out for potholes."

"We will," Samantha said.

Softening and recognizing the inevitable, Kilborn added: "Good luck from me, too. I wish you well with your activities here."

"Thank you, Sir," Samantha said. "*Dylan Evans Base* signing off."
****

At precisely 1655 h, Samantha and Lars interrupted their unloading of supplies and equipment from the rovers. Despite the lower gravity, it had still been hard work lugging everything up the path to the plateau, especially the rover tent. Somewhat exhausted, and welcoming a break, they sat down on a broad rock on the top of the plateau and looked to the east. They held each other's gloved hand in anticipation. At precisely 1700 h, they saw a steak of light climbing up into the sky. The *Hermes* had launched and was heading home. They were the only humans left on Mercury. Both sighed and resumed their activities.

Mercury, Day 88

With aching joints, Samantha and Lars awakened in the reconfigured rover tent the morning of the 88th day. They had spent the previous 2 days setting up their base, then loading *Rover 2* with supplies for their departure to the ice crater later in the morning. Since both rovers had been parked at the bottom of the hill, there was plenty of room for them in the tent. Although they were using some of the oxygen canisters they had brought with them from the *Hermes*, the air supply generator was now operating and was starting to produce more oxygen from the carbon dioxide located outside in the cave. The EM generator and EEG machine were stored until the next Mercury day.

After breakfast, they put on their space suits and went out. They walked along the tunnel to the elevator ladder and ascended. They walked on to the plateau after resealing the opening.

The network rose before them, silhouetted against the sky in the dim light. It loomed like a silent sentinel, guarding their new home. They moved around it to the other side of the plateau. Since they were looking west across most of the relatively flat Caloris Basin, the horizon was easily delineated, with just a few hills and crater ridges rising here and there. The Sun grazed the horizon. Due to the absence of an atmosphere to refract the light, true dusk did not exist, and the sky remained black.

They watched, enraptured, for nearly half an hour, holding gloved hands and thinking private thoughts as the Sun disappeared, signaling the end of another long day on Mercury. Then they turned back, paused for a moment to gaze at the outline of the network, now in darkness, then headed for the path leading down to *Rover 2* and their next adventure.

Part II

The Science Behind The Fiction

A map of Mercury, produced by E.M. Antoniadi around 1920. Note the features that he depicted, which were based on the erroneous premise that Mercury always keeps the same hemisphere facing the Sun (and the back continuously facing the Earth). Courtesy of: *Solar System Maps: From Antiquity to the Space Age,* Nick Kanas, Springer/Praxis, 2014; and Wikimedia Commons. The original digital image was created by NASA and is in the public domain.

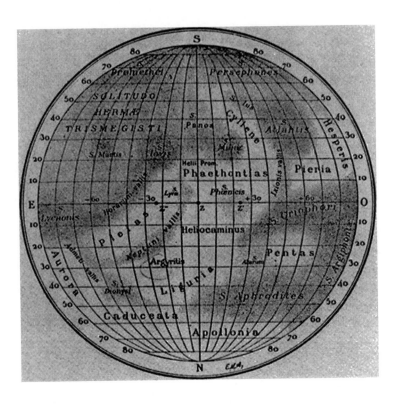

Silicon-Based Life and the Planet Mercury: Fiction and Fact

The science fiction story in this book deals with a silicon-based life form that develops on the planet Mercury. Is this possible? If so, what is the evidence for its occurrence? Most of this section will examine the issue of silicon life on Mercury from a scientific and technical perspective. But first, let's take a look at what has been portrayed historically in the science fiction literature.

Mercury, Silicon Life, and Science Fiction

There have been several stories about Mercury, but none of these have included native life forms. In Alan E. Nourse's short story entitled "Brightside Crossing," first published in 1956 in *Galaxy Magazine* and later reprinted in the book *Tiger by the Tail* [1], a man relates a tale to another explorer of his failed attempt to cross the scalding surface of Mercury during its closest approach to the Sun (perihelion). The attempt ended poorly, with his three colleagues losing their lives and him returning to base. The story is an exciting adventure tale where the conditions of the planet's surface are described in great detail. These include blistering heat, fragile terrain with brittle zinc ledges, and pools of molten lead and sulfur. But the story shows its age, in that in the 1950s Mercury was thought to have one side permanently facing the Sun (the "brightside" of the title) and the other side permanently facing away toward the heavens. As we shall see below, Mercury rotates in a complicated fashion during its revolution around the Sun. Other factual problems include the perception that Mercury is the hottest planet in the Solar System, that winds blow across the surface carrying volcanic ash and sulfurous gases from place to place, and that the planet Venus has a jungle. Nevertheless, the story is well-written and presents interesting psychological issues involving the

© Springer International Publishing Switzerland 2016
N. Kanas, *The Caloris Network*, Science and Fiction, DOI 10.1007/978-3-319-30579-0_2

interactions of the four explorers as they make their way across the dangerous terrain.

In Ben Bova's *Mercury* [2], the planet is portrayed fairly accurately, since the novel was written in 2005, well after the Mariner 10 probe had explored the planet up close and personal. The periods of its rotation and revolution are correctly described, as are characteristics of its surface. But about half of the storyline takes place off-planet: in Ecuador (where a giant space elevator is being built), in space, or on the Moon. The story deals with several people who are interconnected via a terrible accident involving the elevator, and the plot is full of intrigue and revenge. Mercury is portrayed as a hot, barren, desolate planet, full of craters, cracks and ravines, old lava fields, fault lines, mountains, rubble, and dust. There are no native life forms, although Mars, Venus, Jupiter, and its moons have primitive life, and Mars once had intelligent life. Interestingly, Bova alludes to the fact that on hot Venus there are large multi-cellular snake-like organisms consisting largely of silicones that use liquid sulfur as an energy-transfer medium, akin to blood.

In his book *Venus*, written several years earlier [3], Bova gives a bit more information about these life forms. The plot of this book involves a group of people traveling to the planet Venus in order to search for the remains of the first crew to explore the planet, which likely perished. As the searchers float down toward the surface through the thick sulfuric acid clouds, they encounter microscopic, multi-cellular ciliated "aerobacteria" living in the soupy material that begin to eat the metal of their ship and the sealant around its airlock hatches. One of the scientists concludes that the organisms need metal as trace elements, the way we need vitamins. The crewmembers escape destruction of their ship by diving below the cloud bank. On the barren, desolate surface, which can be seen to glow at night from the heat like smoldering charcoal, they locate the wreckage of the previous expedition. Around the wreckage, one of the crewmembers in a manned scout ship encounters long tentacle-like structures that are found to be feeding tubes of a giant underground organism that is feasting on the metal remains. From a piece of a tube that attached to the returning ship, the crewmembers find that the entity's outer shell is made of silicone that is strong, flexible, and heat-resistant. Its interior contains complex sulfur compounds, some of which have never been seen before. The entity does not appear to be intelligent, and no details are given about how it can exist in the scalding, hellish environment of Venus' surface. Bova passes over these details by having one of the scientists admit that the silicon entity functions on a totally new kind of chemistry.

As seen by the above examples, Mercury generally has not been described as a proper home for native life. However, it has been used as a base for humans. For example, in David Brin's *Sundiver*, first published in 1980 [4], some of the action takes place under the surface of Mercury, where spherical sunships are based that periodically leave to explore the chromosphere of the Sun. Brin's universe in the year 2246 teems with life forms, generally divided into oxygen-breathers and hydrogen-breathers. In the former category, we meet organisms looking like lizards, teddy bears, bug-eyed bipeds, and indoor shrubs. With the possible exception of human beings, all life forms have advanced through the mentorship of more evolved species, to which they are indebted. Even humans have helped monkeys and dolphins progress to the point where they can communicate with us and even think logically. The plot line concerns the discovery of life existing in the chromosphere of the Sun. Three entities are described: one greenish and toroidal in shape that spins in its fiery environment and travels with others of its kind in herds; a second smaller and bluish shimmering entity whose function seems to be to shepherd the toroidal forms around; and a third more malevolent entity that has a bluish, ghostly appearance that sometimes takes a human-like form and warns the sundivers to keep away. Complicating matters is the attempt of some of the other species in the universe to sabotage these missions in order to keep humans in their place and discourage their independence. However, none of the entities described in this creative novel seem to be silicon-based, and no life forms are described that live on the surface of Mercury. The Sun-facing surface is described as a bleak, inhospitable place, gouged with craters and rills. Sharp shadows cast by mountains are seen, and there are rocks possessing unusual minerals formed by the Sun's heat and radiation, along with strange powders and crystals.

Mercury also has been visualized as a base for non-human, heat-seeking aliens. In Hal Clement's *Iceworld* [5], originally appearing in 1953 as a serial in *Astounding Science Fiction* magazine, an alien drug smuggler and his crew have arrived in our Solar System in a faster-than-light space ship from the planet Saar. They have been trading platinum and other valuable metals with a family on Earth for tobacco, which is highly addictive to them. Their home planet orbits a blue-white sun and is smaller than Earth, but it has temperatures that exceed 500 °C. Being accustomed to the heat, the Saarians view the Earth as a strange "iceworld," where their accustomed liquids are solids and where they need special heat suits to keep warm on the surface. Although they have established a base in a valley on the hot side of Mercury, they need to construct iron mirrors on the valley walls to concentrate sunlight to warm themselves up even more! Mercury is described as a baked, dry lifeless planet, in contrast to Saar, which has a sulfur-based atmosphere; "water" made out of liquid compounds of copper, lead, and sulfur; strong winds; and, at the poles, sulfuric rain. Life on Saar includes crystalline plants that (like the intelligent Saarians) make all the liquid they need in their own tissues. The Saarians are about a quarter smaller than adult humans and breathe gaseous sulfur. They have four tentacles for arms, two double-jointed legs, two independently moveable eyes, no nose, and a broad thin-lipped mouth. It is not explicitly stated that the Saarians themselves have a silicon-based chemistry, and their ambulatory gait and mobile tentacles suggest otherwise. In fact, many of their physiological features (such as a stomach and bodily movements) and behavioral characteristics are more anthropomorphic than alien.

But intelligent silicon-based entities have been described in science fiction. For example, robots and self-aware computers using silicon-based chips have formed the core of many science fiction movies and stories [6]. In a *Star Trek* television episode called "The Devil in the Dark," the crewmembers are summoned by human colonists on the planet Janus VI to deal with a creature that has been killing some of their miners. The landing party encounters the alien, which resembles a lump of molten stone that is able to move around and burrow through rock with the aid of a strong acid. Following a Vulcan "mind-meld," crewmember Spock discovers that the creature is intelligent and a member of a race called the Horta. The species die out every 50,000 years, except for one member whose job it is to care for the silicon nodule "eggs" that are scattered throughout the caverns of its home. These eggs will produce the next generation of Horta. When the human colonists inadvertently destroyed some of the eggs through their mining activities, the caregiver responded to protect them. Peace eventually is restored, with the colonists and the Horta developing a strategy of coexisting with each other. The alien in the story is creatively conceived, although it is difficult to see how it can live in an oxygen-based atmosphere and move around the caverns given its silicon-based chemistry.

In Alan Dean Foster's novel *Sentenced to Prism* [7], the protagonist is sent by his company to find out what happened to an earlier exploration party that was sent to ascertain the economic viability of a newly discovered planet, called Prism. On arriving, he discovers that the members of the party have been killed and that the native life is over-running the base. The life forms on Prism consist of a variety of silicon-based, carbon-based, and organosilicate entities competing with each other to survive. Some are energized by the planet's star (converting its light directly into electricity), others by imbibing different life forms for their protein or metal content. Some are organized according to fractal rather than normal geometry and are difficult to see with human eyes. A treacherous but rich world is described, with a plethora or organisms varying in size, shape, structure, and color. In addition, there are intelligent life forms based on silicon chemistry that exist in groups called Associatives and communicate through a form of telepathy. The protagonist's adventures on this world are chronicled, including his intimate contact with the native life and with members of a rival company sent to take over the planet for themselves.

Although the story is rich and adventurous, it is more fantasy than hard science fiction. There really is no scientific or technical explanation as to how these various life forms can exist in an environment with running water, Earth-like atmosphere and temperature, and competing chemistries.

Examples from the novel: None of the stories described above visualize sentient life forms based on silicon that are native to the planet Mercury. The story in this book, *The Caloris Network*, conceptualizes such an entity on Mercury and develops a plausible explanation for its existence, in the context of its interactions with a human crew vising the planet in their space ship *Hermes*. It is reasonable to expect these two life forms to be quite different from one another, and the trick in the storyline is to credibly describe their interactions without anthropomorphizing the alien, as too often is the case in science fiction. The degree of success of this attempt is left to the judgement of the reader.

Now, let's examine the background of the setting and what we know about the planet that is closest to the Sun.

The Planet Mercury

Humans have observed the planet Mercury since ancient times. For the ancient Greeks, it was one of the visible "wandering stars," and its movement in the heavens was modeled according to an Earth-centered cosmological system [8, 9]. Initially, Mercury was viewed as two objects. Its sunset aspect was named Hermes, after the speedy messenger god, likely because of its rapid movement in the sky. At sunrise, it was called Apollo, after the god of the Sun. In the fourth century BC, the Greek astronomer Eudoxus realized that both of these wandering stars were in fact one object. Consequently, the Romans settled on one name, Mercury, their name for the god Hermes.

After Copernicus developed his Sun-centered cosmological system, Mercury was viewed as the innermost planet orbiting the Sun. However, for a period of time in the nineteenth century, Mercury's position was displaced by Vulcan, a hypothetical planet that was thought to orbit even closer to the Sun. This notion was dispelled in the next century, returning Mercury to its innermost status [9].

The first person to record telescopic observations of Mercury's surface was astronomer Johann Schroeter in the late 1700s. His drawings were not well-defined, and the features he depicted were inaccurate. Similarly, astronomer Giovanni Schiaparelli in the late 1800s produced drawings of the planet that also were poorly defined. He calculated the rotational period of Mercury to be 88 days, which was the same as its orbital period. At about the same time,

American astronomer Percival Lowell recorded streaks on Mercury that were similar to those he had seen on Mars [9]. However, he saw these markings as natural cracks on the surface, not canals made by intelligent life, as he had proposed for Mars.

The first real map of Mercury was made by Eugene M. Antoniadi [9]. Born in Constantinople of Greek descent, he subsequently established himself as an excellent observer. He worked with Camille Flammarion at his observatory at Juvisy near Paris, and later he worked at the observatory at Meudon. From 1914 to 1929, Antoniadi made a number of observations of Mercury that seemed to confirm the rotational findings of Schiaparelli. This meant that it always kept the same face toward the Sun and its opposite side turned away from our star. This was thought to be the side always seen from the Earth. Antoniadi produced a map of Mercury's surface that became the accepted map of the planet for nearly 50 years (see image at the beginning of this section). Over time, it included nearly 300 features that he thought were always seen in the same position. He summarized his findings in 1934 in his book *La Planete Mercure*. His work was very influential, and his nomenclature formed the basis of subsequent maps of the planet.

In the 1960s, radar studies of Mercury showed the sidereal rotation of the planet to be 58.6 Earth days, which is about half of its 116-day synodic period (the time that elapses between two successive identical configurations as seen from Earth). As a result, the same region of the surface faces us every time Mercury is best placed for observation from Earth. This led to the conclusion that Antoniadi and others were mistaken in their thinking that the planet's orbital and rotational periods were identical. Furthermore, later space probes flew by Mercury and produced excellent images of the surface, which bore little resemblance to Antoniadi's map.

Mercury's diameter is only 4879.4 km (3031 miles). This is about 40 % less than the diameter of Earth, 40 % larger than our Moon, and smaller than Jupiter's moon Ganymede and Saturn's moon Titan. Its circumference is 15,329 km (9525 miles), making the distance from equator to pole about the same as going from San Francisco to Detroit. The gravity is 38 % that on Earth. Over 70 % of Mercury's mass is due to its large, partly-molten iron core, and the rest is due to rock-forming silicon minerals (primarily silicates) comprising the planet's mantle and crust. Mercury's axial tilt is close to zero degrees, which means that the Sun is barely above the horizon at its north pole. As a result, the inside of many craters located there would never be exposed to the Sun and are subsequently always cold.

Mercury's orbital period is 88 Earth days. It revolves around the Sun at a mean distance of 57,910,000 km (35,983,605 miles), ranging from 46,001,009 km at perihelion to 69,817,445 km at aphelion. Consequently, the Sun appears two to over three times larger than it does from Earth. Although Mercury is not tidally locked to the Sun, its rotational period of 58.6 days is tidally coupled to its 88-day orbital period, so it rotates one and a half times during each orbit. This 3:2 resonance means that a full day on Mercury (from sunrise rise to sunrise) lasts 176 Earth days, with the time from sunrise to sunset taking 88 Earth days. Given Mercury's orbital characteristics, an astronaut on the surface who is standing in Caloris Basin where the Sun is overhead at perihelion would experience an interesting phenomenon during the local day. The Sun would rise in the east, move in a westerly direction, then slow down and pause as it approached its highest point some 40 Earth days later, slowly retrograde easterly for a few days, then continue its westerly motion until it sets. Moving to a viewing site where the Sun is overhead at aphelion would produce an even stranger sight: a double dawn where the Sun rises in the east, then slowly returns and dips below the horizon, and then rises again to continue its westerly motion [10].

The first space probe to scan Mercury in detail was Mariner 10, the last of a family of American spacecraft that also explored Venus and Mars. Mariner 10 first reached Mercury in March 1974 and conducted three flybys of the planet from 1974 to 1975. It photographed features as small as a few tens of meters across and showed the surface to be pocketed by thousands of impact craters and to contain old lava fields from volcanoes. Also observed was a large open plain, the Caloris Basin, which subsequently was determined to be some 1550 km in diameter. Rimming Caloris are hills, mountains (some nearly 2 km high), and ridges, which suggest that the basin was formed over 3.8 billion years ago from the impact of an asteroid or other piece of space debris that was over 100 km in size (see image that introduces the novel section of this book for a view of Caloris). On the opposite side of the planet, there is an area of hilly "chaotic" terrain that may represent shock waves from this colossal impact. Other ridges and escarpments were imaged by Mariner 10 that were believed to be due to expansions and contractions as Mercury's core cooled and shrank over time. The persistence of the craters and other surface features suggests that the planet's surface is very old, with little evidence of recent activity, such as tectonic movement, to modify it. Mercury also was found to have a magnetic field, suggesting that it has a rotating liquid core that is relatively large in relation to the small size of the planet.

Mariner 10 found that Mercury has essentially no atmosphere to hold the heat generated by the Sun. This causes its temperatures to vary greatly. Although the mean surface temperature on the planet is 179 °C, the minimum can drop to −173 °C at night on the side away from the Sun, and the maximum can reach 427 °C at high noon on the side facing the Sun.

More recently, Mercury was visited by NASA's MESSENGER space probe [11–14]. The name is an acronym for MErcury Surface, Space ENvironment, GEochemistry and Ranging. Launched on August 3, 2004, it made three flybys of the planet in 2008 and 2009 (plus two more that went by Venus). It began orbiting Mercury on March 17, 2011, and continued to study the planet. It extended the mapping of the surface beyond that obtained by Mariner 10. It also provided evidence for past volcanic activity and the presence of a dense iron sulfide layer sandwiched between the mantle and core. The surface material of the planet was found to consist predominantly of iron-poor calcium-magnesium silicates. There was evidence for deposits of water ice near the north pole in the permanent shadows of craters or just under the surface. The ice likely came from comets that impacted the planet in the past. The magnetic field produced by the planet's fluid iron core was found to be offset some 20° to the north of the planet's center. MESSENGER completed its successful mission and crashed onto the planet on April 30, 2015.

Examples from the novel: In the novel, the *Hermes* crewmembers land in Caloris Basin in order to investigate the source of a mysterious radiation "signal" that has been picked up by passing space probes. They find the surface of the planet to be rocky and full of craters and old lava flows. They observe a curious sunrise where the Sun pauses and retrogrades midway in its path at the middle of the long Mercury day. Other features mentioned above are also described in the novel.

Silicon Life on Earth and Mercury

Astrobiologists and other scientists have speculated on ways in which life could evolve on other worlds. In general, what is considered necessary for life is a chemistry based on the reactive element carbon, reasonable temperatures and pressures, and liquid water in which stable life-supporting molecules can form. Life also needs energy and a strategy for replicating itself through generations. Good arguments have been put forth for the existence of life in such exotic environments as beneath the frozen desert or in sheltered hot springs on Mars, in a watery subsurface on Europa, in liquid carbon-based methane or ethane habitats on Titan, and even in distant exoplanets orbiting other stars [15–18]. Even on Earth, there are a number of extremophilic microorganisms that can survive under inhospitable conditions of temperature, radiation, acidity/alkalinity, and pressure [19].

In science fiction, aliens have been depicted in a variety of ways, some more plausible than others [17, 18, 20]. Most of these are variations of carbon-based entities needing an environment dependent on liquid water. However, an interesting and plausible variation would be to consider a world whose life forms are based on silicon [21], an element in the same column in the periodic table as carbon that has the versatility to support its own biochemical niche. On Earth, silicon compounds are plentiful, comprising most of the Earth's crust. In addition, organisms such as plants, marine diatoms, sea urchins, and sand dollars have structural elements that utilize silicon. The problem is that a world capable of using stable silicon-based compounds as the essence of life would need to be either very cold with no oxygen and a liquid other than water to support silanes (which are molecules of silicon and hydrogen), or a world so hot that the only liquid present would be molten minerals to support silicones and silicates (which are molecules of silicon and oxygen). At moderate Earth-like temperatures, silicon compounds involving hydrogen have weak bonds and fall apart, especially in water. In contrast, silicon compounds involving oxygen, such as the silicates found in minerals such as sand, rocks, and clay, often crystallize and involve bonds that are too strong. This makes them incapable of the kind of interactive chemistry in moderate temperatures that are characteristic of carbon compounds [16, 21].

Other than mobile self-aware robots that are able to repair and manufacture more of themselves, it is difficult to envision silicon-based entities that are intelligent and self-replicating. We have no examples on Earth to point to, since carbon-based chemistry is the basis of life and water and oxygen are plentiful. But what about simpler silicon-based organisms? Could these occur? If so, would they be able to replicate themselves? Quite possibly. In fact, there is a notion that our complicated carbon-based life forms have evolved as a result of interactions with silicon-containing clay material that took place eons ago [16, 22].

This has given rise to the so-called clay hypothesis, which has stemmed from the work of biochemist A. G. Cairns-Smith. According to this model, self-replicating silicon clay crystals in solution provide an intermediate step between biologically inert matter and carbon-based organic life on Earth [22]. His argument goes like this. Broadly speaking, organisms are entities that take part in evolution. The very first organisms were inorganic-crystalline in nature, not organic-molecular. They acted like crystal genes, in that they conveyed information that could be passed along through time. For example, twinning is a defect in a crystal lattice developing in a solution where different regions are misaligned, possibly reflecting changing external conditions. As the crystal grows, this defect can be copied layer by layer, in a sense like a mutated gene in our own cells can be replicated in daughter cells. If the defect has an advantage for the crystal, such as allowing it to grow faster and then break up into smaller particles that can be more easily dispersed by the wind or the water, then a sort of natural selection occurs where this crystal becomes more predominant than other crystals which do not have this growth/distribution advantage. Cairns-Smith uses as his model the silicate crystals found in clay. In supersaturated aqueous solutions, they can grow and precipitate out, especially when seeded by a piece of the crystal that is to be formed. In less saturated solutions, the silicate crystals dissolve and can be carried to other places, where again they precipitate out as the crystals accumulate or the water around them evaporates.

But this is not the end of the story. The structure imposed by crystal lattices in clay may have attracted and helped shape the structure of evolving carbon-based molecules, whose versatility in water under moderate temperatures led them to form more complicated long-chained compounds. The characteristics of the silicate lattice may also have influenced other characteristics of these developing compounds [16, 22]. For example, the structure of long carbon chains forming in a soupy rock pool might have reflected the "handedness" of the silica substrate [21]. Those traits found to be most adaptable could later be replicated as carbon-based life became dominant and developed methods of reproducing itself through a genetic takeover.

It should be noted that these notions of Cairns-Smith suggested a sort of pre-stage between the mineral world and life. However, many of the processes in life forms we are familiar with have a parallel with the process occurring in his clay models. These include stable molecular structures, growth, movement, replication of characteristics, and the occasional mutation that can be passed on to "daughter" layers or distributed elsewhere to become the dominant clay crystalline pattern. Thus, one could plausibly defend the notion that many characteristics of life can exist in a chemistry based on something other than carbon, e.g., silicon.

But given this premise, what about the possibility of silicon life on Mercury? Up until now, we have considered crystal growth and development as occurring in an aqueous solution. As mentioned earlier, there is evidence that Mercury contains water ice in shadowed craters at its poles, where it is very cold. However, there is no evidence for the presence of liquid water anywhere on Mercury, and indeed the daytime temperatures on most of its surface can exceed 400 °C. So how might silicon life develop and grow on this planet?

One possibility is through something like the Czochralski process [23–25], discovered by Polish scientist Jan Czochralski in 1916. It is a method for growing single crystal ingots called boules that has found a commercial application in the production of silicon that is used in the semiconductor industry on Earth. In this process, silicon powder is melted down in a crucible that is made of an inert substance like quartz and located in an inert atmosphere that contains gases such as argon. The electrical characteristics of the silicon can be controlled by adding a dopant like phosphorus or boron to the silicon before it is melted. A small seed crystal mounted at the end of a rotating shaft is then lowered into the molten silicon. As it is slowly drawn up, a cylindrical silicon crystal boule forms around it. Depending upon how much silicon there is in the crucible, the boule can be up to a foot or so wide and several feet long. Subsequently, thin wafers can be cut from these ingots and polished for making integrated circuits or textured for making solar cells. Note that this process is not necessarily related to life. However, it does provide one example of how silicon crystals can grow in the absence of water given the right combination of circumstances.

Could a Czochralski-like process occur naturally on Mercury that would allow silicon crystals to grow? Possibly, but only under certain circumstances. Mercury has ample silicate material in its mantle and on the surface on the planet. The melting point of silicon and many of its compounds is over three times the maximum surface temperature found on the planet, so some form of energy would be necessary to raise the temperature. One way would be to concentrate the energy from the relatively nearby Sun via crystalline refraction (such as might occur on Earth using a magnifying glass), or by tapping into the heat found in the molten core of the planet via a surface vent. We know that Mercury has a molten iron core that is relatively close to the surface. This has resulted in volcanic activity in the past, and the movement of this core gives the planet a magnetic field.

Should a silicon crystal-containing asteroid crash on the planet in an area near a thermal volcanic vent, it might be possible for the crystals to take hold and grow. During the local day, the heat from the Sun would add to that from the vent, and this could cause the surrounding silicon-containing rocks to melt, producing ionic silicon. As this area cooled during the local night, crystals could form around the silicon seed crystals from the asteroid, and the mass could grow in size. Such a silicon mass could be said to have many of the characteristics of life: it uses solar and underground heat energy to sustain itself and grow; the resulting crystals would reproduce the pattern characteristic of the "parent" mass; and any structural changes in this pattern ("mutations") would also be replicated in the newly-formed crystalline lattice.

Examples from the novel: In the novel, the *Hermes* crew discovers an alien life form living on Mercury. It is composed of silicon rods connected by fibers to form a crystalline network. Concentrating the heat generated from the Sun and from volcanic material uplifting from below the surface, the network is able to grow by producing molten silicon ions from rocks buried deep in the regolith during the hot daytime. This material crystallizes out around the existing rods and adds to the lattice during the nighttime when the temperature cools down. Repairs to the network can also be made by the same process.

But that's not all. The crewmembers soon discover that the network is a conscious entity.

Could Conscious Silicon-Based Life Exist on Mercury?

How could such a mass possess consciousness? To examine this possibility, we should first note that thermally and gravitationally driven molten fluids that contain silicon and other ions can generate electromagnetic (EM) waves, and in some cases the frequency of these waves can be less than 100 Hz (cycles per second) [26–29]. This is the sub-radio, or "extremely low frequency" (ELF), range of the EM spectrum. Interestingly, the endogenous waves that are produced by the billions of neurons in our brains are also in the ELF range, generally 1–45 Hz. With this in mind, let's examine electromagnetic theories of consciousness.

New Zealand neurophysiologist Susan Pockett has proposed a hypothesis that consciousness is identical to certain spatiotemporal patterns in the EM field, primarily in the ELF range of the spectrum [30]. She identifies three types of consciousness: simple sensory experiences and perceptions, awareness of the self as an individual entity, and the experience that we are all part of an all-encompassing Self or God. In terms of simple consciousness, she has reviewed a number of studies involving humans and animals that support the existence of specific spatiotemporal patterns in an EM field that covary with: (1) states of consciousness, such as waking, sleeping, and dreaming; (2) non-intentional states, such as what is produced by meditation; and (3) sensory experiences, such as olfaction, hearing, and seeing. She distinguishes her notion that consciousness can result from specific patterns in any EM field from the more traditional model espoused by other neuroscientists that consciousness is produced by the firing of a vast assemblage of neurons and their associated chemicals. In fact, her hypothesis raises the possibility that consciousness can occur in non-neuronal settings as well, such as in artificial intelligence systems or in a god-like universal mind.

Anglo-Irish biochemist and geneticist Johnjoe McFadden also has proposed a model of consciousness based on electromagnetic waves. He has argued that the brain's endogenous EM field is related to the information produced by the neurons in the brain and serves to pool and integrate this information. Furthermore, he has cited evidence from numerous human and animal studies that the synchronicity, rather than the number, of neuronal firings correlates with experimental evidence of awareness and perception [31, 32]. He feels that synchronicity serves to phase-lock EM field fluctuations, thereby increasing their magnitude and influence. He proposes that the EM field is the physical substrate of consciousness and awareness, and that conscious volition is the component of this field that is transmitted back to the neurons and is deduced through its effect on our motor neurons. He has put these ideas into what he calls the conscious electromagnetic information (or CEMI) field theory [31, 32], and he has defended his theory against a number of objections proposed by others [33]. Although containing many ideas similar to those advocated by Pockett, McFadden pays less attention to issues involving the universal mind and focuses more attention to the link between an individual's consciousness and its relationship to neurons.

Brain waves are relatively weak in intensity. On Earth, much stronger ELF waves are generated naturally by earthquakes and lightning and artificially as a by-product of devises such as electrical transmission lines and household appliances [34]. These waves are able to propagate significant distances and even travel through rock and sea water. For the latter reason, they have been useful in communicating with submarines. A space probe has also detected ELF waves during its descent toward the surface of the moon Titan, possibly due to an interaction with Saturn's magnetosphere or related to intrinsic fields on the moon itself [35].

Given enough intensity, ELF waves can have effects on biological systems. The best known effects are thermal, but evidence is mounting that non-thermal effects also can occur. For example, there have been studies where human subjects have perceived tingling or painful effects of ELF waves in the 50–60 Hz range [36, 37], and both visual-perceptual and motor coordination effects have been reported when subjects were placed in an electrical field applied to the head whose stimulus frequency was characteristic for the sensory or motor cortical brain region being tested (generally in the 10–20 Hz range) [37]. Similarly, statistically significant changes in the social behavior of baboons have been found when they were placed in a 60 Hz electric field measuring 30 and 60 kV/m [38, 39].

All this suggests that the movement of a molten mass of silicon ions associated with a silicon crystal lattice could produce low frequency electromagnetic waves that are in the same frequency range as brain waves, but much more intense. Although such ELF waves would be caused by natural physical processes, it is possible that they could result in a kind of primitive consciousness. Of course, the likelihood of all this occurring on a relatively inactive planet such as Mercury is low, especially since it requires a complex set of circumstances and billions of years. But self-replicating, carbon-based intelligent life on Earth also developed from a complex series of events over billions of years. Who says that a similar phenomenon couldn't happen on a planet like Mercury that is silicon-based?

Examples from the novel: Since Mercury has a gravitational field, the hot moving silicon ions produced by the network generate extremely low frequency waves akin to human brain waves that can traverse its crystalline rods and fibers. Consequently, the network in this story has developed a kind of consciousness, albeit a very primitive one. It perceives itself as an entity, and it conceptualizes the Sun as another entity and its creator, since the network is only conscious when the Sun is up and since the periodic rising and setting of our star is the primary stimulation it receives from its environment. When threatened, the network reacts in a hyperaroused manner, producing intense ELF waves that can be concentrated and directed toward the offending agent, destroying it in the intense heat. This process reaches a point where the radiation begins to damage the *Hermes*, thus preventing the crewmembers from launching their ship to escape. When the crew realizes what is happening, they try to communicate with the network as one conscious species to another so that they can survive.

References

1. Nourse, A.E.: Tiger by the Tail and Other Science Fiction Stories. MacFadden Books, New York (1964)
2. Bova, B.: Mercury. Tor Books, New York (2005)
3. Bova, B.: Venus. Tor Books (Mass Market Edition), 2001.
4. Brin, D.: Sundiver. Bantam Books, New York (1995)
5. Clement, H.: Iceworld. Ballantine Books, New York (1981)
6. Lerner, E.M.: Alien aliens: beyond rubber suits. Analog Science Fiction and Fact, April 2013, 39–48 (2013)
7. Foster, A.D.: Sentenced to Prism. Ballantine Books, New York (1985)
8. Kanas, N.: Star Maps: History, Artistry, and Cartography, 2nd edn. Springer, New York (2012)
9. Kanas, N.: Solar System Maps: From Antiquity to the Space Age. Springer, New York (2014)
10. For details and an animation showing daily temperatures and the path of the Sun as viewed from within the Caloris Basin, see the following MESSENGER website. http://www.messengereducation.org/Interactives/ANIMATIONS/Day_On_Mercury/day_on_mercury_full.htm ("A Day on Mercury," courtesy of NASA, in conjunction with the Carnegie Institute for Science, and John Hopkins Applied Physics Laboratory)
11. Solomon, S.C. McNutt, R.L., Jr., Anderson, B.J., Blewett, D.T., Evans, L.G., et al.: Messenger's Three Flybys of Mercury: An Emerging View of the Innermost Planet.

41st Lunar and Planetary Science Conference, The Woodlands, Texas, March 1–5, 2010. http://www.lpi.usra.edu/meetings/lpsc2010/pdf/1343.pdf

12. Bedini, P.D., Solomon, S.C., Finnegan, E.J., Calloway, A.B., Ensor, S.L., et al.: Messenger at mercury: a mid-term report. Acta Astronaut. **81**, 369–379 (2012)

13. Oberg, J.: Torrid Mercury's icy poles. Astronomy, December 2013, 30–35 (2013)

14. MESSENGER mission website. http://messenger.jhuapl.edu/. Accessed 1 Oct 2014

15. Nott, J.: Titan: a distant but enticing destination for human visitors. Aviat. Space Environ. Med. **80**, 900–901 (2009)

16. Irwin, L.N., Schulze-Makuch, D.: Cosmic Biology: How Life Could Evolve on Other Worlds. Springer, New York (2011)

17. Kanas, N.: The New Martians: A Scientific Novel. Springer, New York (2014)

18. Kanas, N.: The Protos Mandate: A Scientific Novel. Springer, New York (2014)

19. Dartnell, L.: Biological constraints on habitability. Astron. & Geophys. **52**(1), 25–28 (2011)

20. Baxter, S.: Imagining the alien: the portrayal of extraterrestrial intelligence and SETI in science fiction. Br. Interplanet. Soc. **62**, 131–138 (2009)

21. Dessy, R.: Could silicon be the basis for alien life forms, just as carbon is on Earth? Scientific American. http://www.scientificamerican.com/article/cfm?id=could-silicon-be-the-basi&print=true. Accessed 23 Feb 1998

22. Cairns-Smith, A.G.: Seven Clues to the Origin of Life. Cambridge University Press, Cambridge (1985/1991)

23. Top-Alternative-Energy-Sources.com.: The Czochralski Process. http://www.top-alternative-energy-sources.com/Czochralski-process.html. Accessed 22 Feb 2015

24. Wikipedia.: Czochralski Process. http://en.wikipedia.org/wiki/Czochralski_process. Accessed 22 Feb 2015

25. Talik, E., Oboz, M.: Czochralski method for crystal growth of reactive intermetallics. Acta Phys. Pol. A. **124**, 340–343 (2013)

26. Johnston, M.J.S.: Review of magnetic and electric field effects near active faults and volcanoes in the U.S.A. Phys. Earth Planet. Inter. **57**, 47–63 (1989). https://profile.usgs.gov/myscience/upload_folder/ci2010Nov231351564287193.pdf. Accessed 18 Apr 2015

27. Karakelian, D., Klemperer, S.L., Fraser-Smith, A.C., Beroza, G.C.: A transportable system for monitoring ultra-low frequency electromagnetic signals associated with earthquakes. Seismol. Res. Lett., **71**(4), 423–436, July/August 2000. https://www.quakefinder.com/research/EQTdata/Karakelian-2.pdf. Accessed 18 Apr 2015

28. Johnston, M.J.S.: Ch. 38. Electromagnetic fields generated by earthquakes. Int. Handb. Earthq. Eng. Seismol. **81A** (2002). https://profile.usgs.gov/myscience/upload_folder/ci2010Nov2222285142871139.pdf. Accessed 18 Apr 2015

29. Grahalatshmi, M.: U.S. Geological Survey Professional Pages. V. Volcano-electromagnetic effects. Compiled November 11, 2006. https://profile.usgs.gov/myscience/upload_folder/ci2010Nov2221320042871158.pdf. Accessed 18 Apr 2015

30. Pockett, S.: The Nature of Consciousness: A Hypothesis. Writers Club Press, Lincoln, NE (2000)

31. McFadden, J.: Synchronous firing and its influence on the brain's electromagnetic field: evidence for an electromagnetic field theory of consciousness. J. Consc. Stud. **9** (4), 23–50 (2002). http://philpapers.org/archive/MCFSFA.pdf. Accessed 16 Apr 2015

32. McFadden, J.: The CEMI field theory: closing the loop. J. Consc. Stud. **20**(1–2), 153–168 (2013). http://epubs.surrey.ac.uk/763034/1/mcfadden_JCS_2013%28a %29.pdf. Accessed 16 Apr 2015

33. McFadden, J.: The conscious electromagnetic information (Cemi) field theory: the hard problem made easy? J. Consc. Stud. **9**(8), 45–60 (2002). http://philpapers.org/ archive/MCFTCE.pdf. Accessed 16 Apr 2015

34. Raz, A.: Could certain frequencies of electromagnetic waves or radiation interfere with brain function? Mind and brain: ask the experts. Scientific American, April 24, 2006. http://www.scientificamerican.com/article/could-certain-frequencies/. Accessed 13 Apr 2015

35. NASA.: Cassini: unlocking Saturn's secrets. Titan's mysterious radio wave, June 1, 2007. http://www.nasa.gov/mission_pages/cassini/whycassini/cassinif-20070601-02.htm. Accessed 13 Feb 2015

36. World Health Organization.: Environmental Health Criteria 238. Extremely Low Frequency Fields (2007), pp. 1–14. www.who.int/peh-emf/publications/elf_ehc/ en/. Accessed 13 Apr 2015

37. International Commission on Non-Ionizing Radiation Protection.: ICNIRP Guidelines for limiting exposure to time-varying electric and magnetic fields (1 HZ to 100 kHz). Health Phys. **99**(6), 818–836 (2010). http://www.icnirp.org/cms/ upload/ publications/ICNIRPLFgdl.pdf. Accessed 13 Apr 2015

38. Coelho, A.M., Jr., Easley, S.P., Rogers, W.R.: Effects of exposure to 30 kV/m, 60-Hz electric fields on the social behavior of baboons. Bioelectromag. **12**(2), 117–135 (1991). http://www.ncbi.nlm.nih.gov/pubmed/2039556. Accesssed 24 Apr 15

39. Easley, S.P., Coelho, A.M., Jr., Rogers, W.R.: Effects of exposure to a 60-kV/m, 60-Hz electric field on the social behavior of baboons. Bioelectromag. **12**(6), 361–375 (1991). http://www.ncbi.nlm.nih.gov/pubmed/1750829. Accessed 24 Apr 2015

CPSIA information can be obtained
at www.ICGtesting.com
Printed in the USA
LVOW10s2059190717

541902LV00006B/126/P